TRANSISTOR CIRCUIT ACTION
SECOND EDITION

HENRY C. VEATCH

Instructor and Electronics Coordinator, Retired
San Leandro Adult School

Editor in Chief
Technical Writing Associates
Reno, Nevada

Gregg Division
McGraw-Hill Book Company

New York   St. Louis   Dallas   San Francisco
Auckland   Bogotá   Düsseldorf   Johannesburg
London   Madrid   Mexico   Montreal   New Delhi   Panama
Paris   São Paulo   Singapore   Sydney   Tokyo   Toronto

Library of Congress Cataloging in Publication Data

Veatch, Henry C
   Transistor circuit action.

   Includes index.
   1. Transistor circuits. I. Title.
TK7871.9.V4 1977   621.3815'3'0422   76-859
ISBN 0-07-067383-7

TRANSISTOR CIRCUIT ACTION

3 4 5 6 7 8 9 0 DODO   8 3 2 1 0

The editors for this book were Gordon Rockmaker
and Susan L. Schwartz, the designer was Victoria
Wong, the art supervisor was George T. Resch, and
the production supervisor was Regina R. Malone. It
was set in Baskerville by Kingsport Press, Inc.
Printed and bound by R. R. Donnelley & Sons
Company.

CONTENTS

PREFACE

Rapidly advancing technology has made a second edition of *Transistor Circuit Action* necessary. New and revised material has been added, but the book retains the simplicity that many users found desirable in the first edition. The added material encompasses more extensive diode coverage, newer types of transistors (silicon-epitaxial planar, for example), a mathematical derivation of the ERCA equations, a wider variety of circuits and applications, and an introductory chapter on linear integrated circuits. The changes reflect many constructive suggestions from users to make *Transistor Circuit Action* even more useful, up-to-date, and complete. Basically, however, the book retains its simple and entirely practical approach to the study of linear transistor circuits.

Several areas of interest have been strengthened. More practice problems are presented at the end of selected chapters, and they appear in ascending order of difficulty. Many of the existing problems have been reworked to make them even more interesting and informative, and to make the grading simpler. A wide variety of circuit applications has been added, such as differential amplifiers, operational amplifiers, and several more oscillator circuits, among others.

Another subject that has been given more attention is the equivalent-circuit approach to describing transistor operation. Several equivalent circuits of varying degrees of complexity are shown. In circuit analysis problems, however, the approach remains to use the simplest one that will yield the desired result, as was the case in the earlier edition.

A very real effort has been made to increase the amount of material relating to the signal activity of a transistor amplifier. Even in the earlier chapters the concept of an active device is introduced to ensure that students who have no prior experience with electronic equipment can obtain a feeling for the process of amplification.

Finally, because integrated circuits are being used so widely, Chapter 17 serves to familiarize students with this most important subject.

Henry C. Veatch

CHAPTER

1 SEMICONDUCTOR MATERIALS

1-1 ATOMIC STRUCTURE

All matter—the "stuff" of which our physical world is made—consists of minute particles. The granular structure of matter was long suspected but only recently proved. The fundamental building block of all matter is the atom. From the various atoms are constructed the molecules, each with a set of characteristics that is distinctive of the material itself.

As an example, a molecule of water consists of two atoms of hydrogen and one atom of oxygen. These atoms are bound together with relatively strong forces, and under normal circumstances cannot be separated. Evidence of this is the fact that water can be boiled or frozen, but it still remains water. Even in the form of water vapor, an invisible gas, the individual water molecules are unchanged, being identical with any water molecule, no matter what the physical state.

A molecule is usually defined as the smallest particle of a chemical compound that still retains the characteristics of the compound. A compound is any material that consists of more than one kind of atom in chemical bond. In the foregoing example, water is classed as a compound, since each molecule of water consists of both hydrogen and oxygen.

Both hydrogen and oxygen are classed as elements because they cannot be simplified or broken down into similar substances by chemical means. Thus the building blocks of elements are atoms. (Elements can be transmuted, or changed into other elements, by disturbing the nuclear forces within the atom. This, however, is not a chemical process.)

There are 92 naturally occurring elements known. Hence there are 92 different kinds of atoms, since an element can contain only one kind of atom. The simplest, and therefore lightest, atom is hydrogen. This is true because an atom of hydrogen has the fewest possible number of constituent particles. On the other hand, uranium is the heaviest element found on earth. Each atom of uranium contains a large number of particles, each contributing to the weight of the atom.

The individual particles making up a typical atom are called *electrons, protons,* and *neutrons.* It is not presently known what these particles are made of, but we do know that they possess both mass and a force that we call the electrical force. The electron exhibits a negative electrical force, while the proton exhibits a positive electrical force. The neutron has no electrical charge.

A simplified idea of the structure of a typical atom is useful in describing how semiconductor materials function. A drawing of an atom is shown in Fig. 1-1 in simplified form. A central part, called the

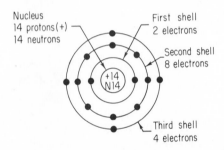

FIG. 1-1. Structure of a silicon atom.

nucleus, contains most of the mass of the atom. The nucleus contains both protons and neutrons (except in the case of hydrogen, which has no neutrons). The electrons circle around the nucleus in somewhat the same manner that the planets circle about the sun.

The electrons form into layers, or rings, or shells. The innermost ring can contain up to two electrons, but no more. If more than two electrons must be accommodated, a second ring is formed which may contain up to a maximum of eight electrons. Again, if necessary, a third and fourth ring are formed to handle large numbers of electrons.

The number of electrons in a given atom is determined by the number of protons in the nucleus. In a neutral atom, there is one electron in one of the rings for every proton in the nucleus. Thus an atom of hydrogen, which has one proton in the nucleus, has one electron circling around the nucleus. An atom of neon has 10 protons in the nucleus, so must have 10 electrons circling about the center.

An orderly listing of the known elements is called a *periodic table* of the elements. The periodic table begins at hydrogen, and lists all 92 elements in increasing order of atomic weight. (Some tables list the artificially produced elements heavier than uranium.) The element number indicates the total number of electrons per atom. For instance, one of the materials used in the manufacture of semiconductors is germanium, element number 32. Germanium has 32 protons in the nucleus and 32 electrons in the various rings.

A portion of the periodic table appears in Fig. 1-2. The elements in this part of the table are those used in the manufacture of semiconductors. Of those shown, the two in greatest use are germanium and silicon. The two, however, will seldom (if ever) be used together. A semiconductor will be made from either one or the other, with very small quantities of one or more of the others used as additives.

III	IV	V
5 Boron	6 Carbon	7 Nitrogen
13 Aluminum	14 Silicon	15 Phosphorus
31 Gallium	32 Germanium	33 Arsenic
49 Indium	50 Tin	51 Antimony
81 Thallium	82 Lead	83 Bismuth

FIG. 1-2. Portion of periodic table.

Used in the manufacture of semiconductors

1-2 VALENCE ELECTRONS

The rings, or orbits, in which the electrons spin appear in a highly organized fashion. A simplified diagram of both a germanium and a silicon atom is shown in Fig. 1-3, along with the more complete representation. The outermost ring of electrons is called the *valence ring*. It is the only ring of electrons that interests us. We shall therefore show only the valence ring in the simplified drawing. The rest of the atom is represented by the inner circle.

The valence electrons influence many of the physical attributes of a given material. Although they are the major item that dictates the chemical activity of a material, we of course are not interested in this phase of their influence. We are interested in the fact that they have a bearing on the electrical conductivity of a substance. Good conductors have one thing in common: they have one, two, or three valence electrons. Hence the outer, or valence, ring of electrons is not completely filled, and this is what makes the material a good electrical conductor. For example, copper has one valence electron, while aluminum has three. Each of these is an excellent conductor of elec-

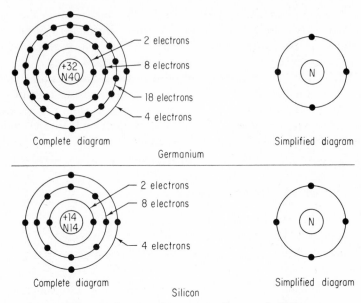

FIG. 1-3. Comparing germanium and silicon.

.trical current. They also easily form compounds with other similar materials.

On the other hand, if the valence ring is completely filled, the material is a very poor conductor. Also, in this case, there is no tendency for the material to enter into a chemical bond with other substances to form a chemical compound. Each electron ring is divided into subshells, as suggested by Fig. 1-4, which represents the element germanium. Each lettered group is a ring, or shell, and all but the first have more than one level. It has been found that any substance that has eight electrons in the outer shell is a stable element. That is,

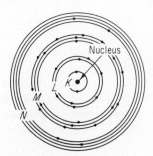

FIG. 1-4. Germanium atom, showing the actual division of electron shells into subshells. Each shell (labeled L, M, N) is divided into subshells; however, the K shell has but one level. The shells extend even farther for the heavier elements.

there is little tendency to combine chemically with other substances. Also, there is little tendency to conduct electrical current.

Referring again to Fig. 1-3, note that the simplified drawing of both germanium and silicon indicates that each has four valence electrons. Recalling that good conductors have few valence electrons while poor conductors have eight electrons in the valence shell, we can appreciate why germanium and silicon are called semiconductors. Electrically speaking, these two substances fall about halfway between the good conductors and the good insulators.

1-3 CRYSTALLINE SUBSTANCES

When the atoms of a material are separated by a distance greater than the diameter of one atom, there is only a slight attraction between atoms. Such a substance is a gas, or a mixture of gases, such as air. When the atoms are separated by a distance about equal to the diameter of one atom, the material is a liquid. When the atoms are so close together that they overlap, the attractive forces become large, and the material is a solid. Each atom is firmly locked in a fixed position. The atoms of most materials, when solidified, form a geometric pattern that is quite uniform. Some substances, however, such as glass or certain plastics, do not form a definite pattern when solidified. Among familiar substances that are examples of crystals are salt and sugar. Most people have viewed grains of salt under low-power magnification and know that the grains are usually in the form of cubes. The atoms of salt are aligned in this same cubical shape. That is, the grains of salt that we can view are simply a manifestation of the configuration of the atoms themselves.

Some materials, such as salt, form crystals whenever they are solidified. Others form crystals only under extreme conditions. Common everyday carbon will crystallize only under the condition that very high temperature and pressures are used. In its crystalline form it is, of course, a diamond.

Germanium and silicon are among those elements that crystallize rather easily. It is in their crystalline form that they perform so well as semiconductors. A simplified representation of germanium appears in Fig. 1-5. Note the geometrical symmetry that exists throughout the entire structure. The bonding between the atoms is shown by the dashed lines, labeled the "covalent bond." This represents the attractive force that holds the atoms locked into the crystal formation. The attractive force is due to the fact that the atomic structure of a solid requires that the atoms have eight electrons in the outer ring if at all possible. In order to satisfy this requirement for the atoms shown in Fig. 1-5, each atom must "share" its electrons with the neighboring atoms. Thus, although each atom actually has only four electrons in the valence ring, the effect is the same as though it had eight. That is, because the valence rings overlap, each atom acts as if it had a full outer ring. Thus each electron is locked in its place, not only giving greater strength to the material (witness the diamond's hardness), but also providing very low electrical conductivity. That this last statement is true is evident from the fact that there are no free electrons

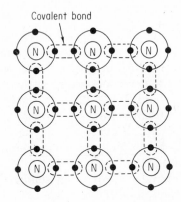

Covalent bond

FIG. 1-5.   The covalent bond of a germanium atom.

available to act as current carriers. The electrons are so tightly bound in their orbits that it requires a large amount of energy to remove them.

Pure or intrinsic crystalline germanium or silicon, then, is a poor electrical conductor. Since it is impossible to purify a material completely, there will always be small amounts of other substances that will cause a small current to flow with a suitable emf applied. An additional fact contributes to the small current that will flow in a pure semiconductor. Temperature has the effect of agitating the electrons. As the temperature is raised, a few of the electrons may be momentarily removed from the covalent bond long enough to act as current carriers. Hence a small current can be caused to flow, the amount depending upon the number of impurity atoms and the temperature of the material. Nevertheless, it is still a poor conductor at room temperatures. The effect of temperature on semiconductors will be dealt with later in detail.

1-4 N-TYPE GERMANIUM

Pure germanium (or silicon) is of little use as a semiconductor. However, by the addition of a small amount of another material, the electrical characteristics are altered enough to be appreciable. If a very
small amount of one of the elements in Fig. 1-2, column V, is added to
the pure germanium, a structural change occurs that increases the
conductivity of the material. Any of these materials—antimony, for
example—has five valence electrons that will try to fit into the crystal
structure of the germanium. As indicated in Fig. 1-6, four of the valence electrons will form covalent bonds with the surrounding ger-

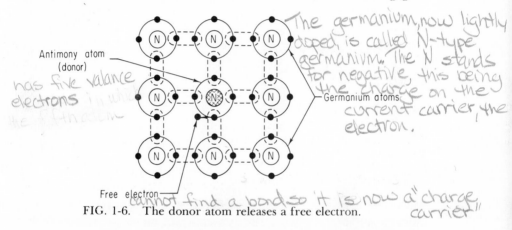

The germanium, now lightly doped, is called N-type germanium. The N stands for negative, this being the charge on the current carrier, the electron.

Antimony atom (donor) — *has five valence electrons in which the fifth electron*

Germanium atoms

Free electron — *cannot find a bond so it is now a "charge carrier"*

FIG. 1-6. The donor atom releases a free electron.

manium atoms. The fifth electron, however, cannot find a bond, and
so becomes a free electron. This free electron is now a "charge
carrier," and will allow the transport of current through the material.
The antimony atom is called the *donor* atom since it contributes, or
donates, the free electron.

   The number of impurity atoms used is very small. About one
atom of the doping agent for every 200 million germanium atoms is
enough to affect very noticeably the characteristics of the material.
The germanium, now lightly doped, is called *N-type* germanium. In
this case, the N stands for negative, this being the charge on the current carrier, the electron.

1-5 P-TYPE GERMANIUM

When one of the substances from Fig. 1-2, column III, is used as a
dopant, a different structure is created. These materials, gallium, for

example, have only three electrons in their valence ring. Again, adding a small amount of gallium to the germanium produces a change in the electrical characteristics of the germanium. Figure 1-7 shows the result of adding gallium to the germanium. The three available electrons form covalent bonds with the surrounding germanium atoms,

Place where an electron would be, called "hole"

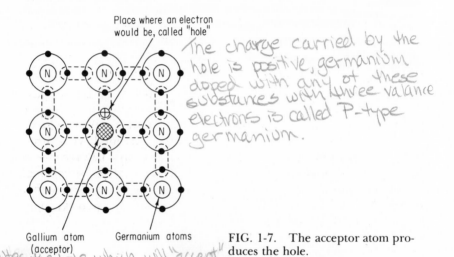

The charge carried by the hole is positive, germanium doped with any of these substances with three valance electrons is called P-type germanium.

Gallium atom (acceptor)   Germanium atoms

contributes the hole which will "accept" the electron

FIG. 1-7. The acceptor atom produces the hole.

but there is one bond that is incomplete. This incomplete bond is called a *hole*, and it has a large attraction for an electron, should one wander by. It is an area in the structure of the material where nothing exists. But since an electron is missing, it acts as if it were a particle carrying a positive charge. The hole, then, does not exist as a "thing." It has no mass, and exists only as a positive charge that will attract an electron if one happens to be available.

The hole, even though it does not exist as matter, nevertheless will contribute to current flow through the material. If free electrons are injected from outside the material, the holes act as a reservoir, allowing the electrons, literally, to jump from hole to hole as they pass through the material. Because the charge carried by the hole is positive, germanium doped with any of these substances with three valence electrons is called *P-type* germanium. The gallium atom, called the *acceptor* atom, contributes the hole which will "accept" the electron, thus giving rise to the name.

Again, the number of impurity atoms used to produce P-type material is very small. About one acceptor atom for every 200 million germanium atoms is used. As before, this changes the electrical conductivity of the material.

1-6 MAJORITY AND MINORITY CARRIERS

The fundamentals of the two types of materials have been discussed up to this point. Still to be discussed is the reaction of the material to some external stimulus.

Take, for a first example, a small quantity of N-type germanium. We know that there exists within the volume of material a certain number of free electrons, called majority carriers, the number depending directly upon the number of donor atoms added. If we were to place an ohmmeter across this piece of N-type germanium, we would find it to have a low resistance, perhaps about 100 Ω (ohms).

It can be seen that the ease with which the current flows depends upon how many free electrons are available to act as current carriers. For a given amount of material this is a function of the number of impurity atoms introduced into the germanium. There is another factor which influences the resistance of the material, upon which we have not as yet touched; this very important factor is the temperature of the material.

At room temperature (70°F, or 21°C) all the donor atoms can be presumed to have contributed to the liberation of free electrons, or majority carriers. Also, at room temperature, *the atoms of the germanium material itself are thermally agitated to the point where a few of the covalent bonds will be broken.* This is most significant, and is pictured in Fig. 1-8. Notice, particularly, that at room temperature there are *free electrons and holes* in the N material produced by thermal agitation. These are produced in both intrinsic and doped materials. However, at room temperature, the number of holes and electrons generated in this manner is very small, compared with the number of impurity-liberated free electrons ("majority carriers"). As the temperature is increased to about 185°F for germanium and 392°F for silicon, the number of covalent bonds that are broken is very large. The conductivity of the material is now determined by the number of holes and electrons created by the high temperature, rather than by the effects of doping. All this simply means that at elevated temperatures the desirable effects of doping completely disappear, and the material fails to perform as a semiconductor. The holes that are thermally generated in the N-type germanium are called *minority carriers.*

The same general effect is apparent in P-type material. At room temperature, each acceptor atom can be expected to have provided a hole (now called the majority carrier) to the volume of material that is available to act as a current carrier. Again, the room temperature will provide sufficient thermal agitation to the germanium atoms so

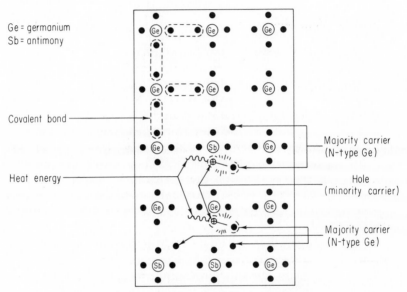

FIG. 1-8.   Temperature effects of N-type Ge.

that an occasional covalent bond is broken, liberating an electron and creating a hole in the place where the electron left. (These electrons are now known as the minority carriers. The hole, of course, becomes a majority carrier in the P material.) The number of these electron-hole pairs created at room temperature is very small compared with the number of holes created by the acceptor atoms.

As the temperature is gradually elevated, the generated electron-hole pairs increase in number very rapidly. When the critical temperature is reached, as before, the electron-hole pairs are far greater in number than the acceptor-atom-produced holes. Now the acceptor-produced holes will have no effect at all on the conductivity, and the P material will fail to perform as a semiconductor at this elevated temperature.

QUESTIONS AND PROBLEMS

True or false:

1-1 Valence electrons are the electrons that take part in a covalent bond. True

1-2 A covalent bond exists when the atoms are loosely bound. False

1-3 Pure germanium is a very good conductor. False

1-4 N-type material is a very poor conductor. False

1-5 P-type material is a very good conductor. False

1-6 Temperature has no effect on semiconductors. False

1-7 The process of doping means to purify. False

1-8 A donor atom liberates a free electron. True

1-9 An acceptor atom has three valence electrons. True

1-10 A hole is a particle that carries a positive charge. True

1-11 N-type germanium carries current because of free electrons. True

1-12 Under some conditions impurity-liberated carriers are insignificant in number.

1-13 There are some conditions where minority carriers become greater in number than majority carriers. False

CHAPTER

2 THE PN JUNCTION DIODE

2-1 THE PN JUNCTION

By themselves, the separate P and N materials are of little practical use to us, as previously stated. If, however, a junction is made, consisting of a piece of P-type material joined to a piece of N-type material, so that the crystal structure is unbroken, a device is produced which *is* extremely useful. We call such a device a *diode*, and its usefulness stems from the fact that it will allow current to flow through it *in one direction only.* The unidirectional properties of a diode allow us to "steer" electrical current into a certain path by allowing the passage of current under certain conditions, and disallowing it under the reverse conditions. We call this process *rectification*, and the device itself we call a *rectifier*, or diode.

How is it possible that, by properly joining two nearly identical pieces of material, each of which, by itself, will freely conduct current in any direction, it now refuses to allow conduction in one direction? The answer to this is, without doubt, one of the more interesting occurrences to be studied in the field of electronics.

Consider, first, the condition of the P- and N-type germanium just prior to joining. (Actually, we cannot just push two pieces together; the germanium atoms at the junction must offer an unbroken path from the N type through to the P type. That is to say, the whole crystal is one complete piece, and during manufacture the crystal is *doped* in alternate layers. When it is cut at the proper place, this will separate the layers so that the junction lies between the cut ends. The description is much simpler, however, if we assume, for the present purpose, that we can join together a small quantity of P- and N-type germanium.) In Fig. 2-1, a small section of N and P material is shown just prior to the formation of the junction. Keep in mind that the majority carriers are in constant motion, as are the minority carriers. The minority carriers are thermally produced, and they exist only a short time, after which they recombine. In the meantime, others have been produced; and this process goes on and on, the number that exist at any one time depending upon the temperature of the material. The number of majority carriers is, however, fixed, depending on the number of impurity atoms available. While the above-mentioned particles are in motion, it is important to realize that

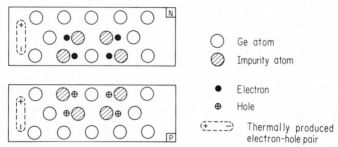

FIG. 2-1.   Semiconductor materials before forming the junction.

the germanium and impurity atoms themselves are fixed in place within the structure of the solid material.

We now turn to the condition of the materials at the instant the junction is formed. (Again, we shall take the liberty of assuming that we can merely push the two pieces together to form the junction.) A completely different set of conditions will now exist. As soon as the junction comes into being, two things occur. The free electrons in the N material "look" across the junction and see a region that has very few free electrons. At the same time the holes in the P material "look" across the junction and see a region that has very few holes. The electrons and holes (majority carriers) begin to diffuse across the junction, i.e., wander across. Then a few electrons appear in the vicinity of the P material, and a few holes appear in the N material. The chances are excellent that the hole and the electron will collide, the negative charge on the electron canceling the positive charge of the hole, and both will cease to exist as charge carriers. After several collisions occur, an electrostatic field exists between the P-type and the N-type material. The source of this field, or potential, is depicted in Fig. 2-2, where only the impurity atoms are shown, plus a few Ge atoms. Keeping in mind the actual preponderance of Ge atoms compared with impurity atoms, let us see what is happening.

The impurity atoms are, of course, fixed in their individual places. The atom itself is part of the solid structure of the crystal and so cannot move about. When the electron and hole meet, their individual charge is canceled, and this leaves the originating impurity atom with a net charge. The atom that produced the electron now lacks an electron, and so becomes charged positively, whereas the atom that produced the hole now lacks a positive charge, and so becomes negative. The electrically charged atoms are called *ions*, since they are no longer neutral.

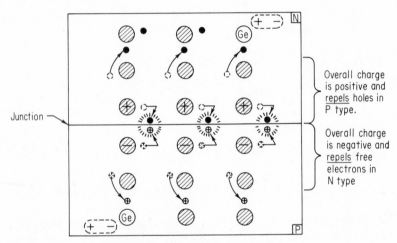

FIG. 2-2.   Forming the junction.

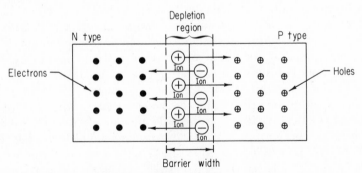

FIG. 2-3.   The depletion region.

After several collisions occur, the field produced by the sum of the individual impurity-atom charges is great enough to repel the rest of the majority carriers away from the junction. So after a time, a condition of equilibrium exists, and the crystal then remains static; nothing further happens. The net result of this field is that it has produced a region, immediately surrounding the junction, that *has no majority carriers.* The majority carriers have been repelled away from the junction, and so are not available as carriers of current. We find that they have been caused to be concentrated nearer the ends of the material, leaving the junction depleted of carriers. The junction is known as the *barrier region,* or depletion region.

If we simplify Fig. 2-2, we can visualize this more clearly. Figure 2-3 clearly shows the lack of carriers in the vicinity of the junction and the large number concentrated away from the junction.

2-2   REVERSE BIAS

The diode that has just been described is now capable of exhibiting the property of rectification. Of course, up to this point, we can only

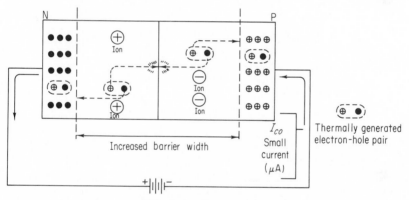

FIG. 2-4.   Reverse bias.

assume that this is true, for no mention has been made of current flow through the device. If we connect our semiconductor to a source of voltage, called "bias" when applied to the diode, we can determine how the device reacts.

Figure 2-4 assumes a certain polarity of applied voltage, or bias, that will put a positive voltage on the N side of the diode, and a negative voltage on the P side. Notice that the majority carriers are attracted by the battery, pulling them *farther away* from the junction. The barrier width has been increased, and when the restraining force of the resultant field within the confines of the barrier region just equals the applied bias, an equilibrium condition is again established. Note that *there is not, nor can there be, current flow* that could be attributed to the majority carriers. It is useful to think of the barrier region as an insulator that will not allow current flow.

The above statement that no current can flow is true if we consider only the majority carriers. However, in Chap. 1, mention was made of thermally generated electron-hole pairs. At some given temperature, some number of these electron-hole pairs is generated throughout the volume of the material. We must consider these current carriers if they are generated in the vicinity of the junction, and if the applied voltage is as shown in Fig. 2-4. The electron-hole pairs are shown by the symbol $(\oplus\ \bullet)$, and the majority carriers by $\bullet$ and $\oplus$.

The negative voltage applied to the diode will tend to attract the hole thus generated and repel the electron. At the same time, the positive applied voltage will attract the electron toward the battery and repel the hole. *The electron in the P material and the hole in the N material are being forced to move toward each other, and will probably combine.* We have removed an electron from the P side and a hole from the N side. When an external voltage is applied, any hole-electron combination in the area of the junction *will cause current to flow in the entire circuit.* An electron from the battery will enter the P material to replace the one lost in the combination, and an electron in the N material will flow out toward the battery. The current that flows, as described above, is due to minority carriers and is usually very small, on the order of a few microamperes. But if the temperature is increased appreciably, the number of minority carriers increases, and the current must increase also.

The bias shown in Fig. 2-4 is called *reverse bias*, since practically no current exists at normal temperatures. Therefore the small current that does flow is called cut-off, or "reverse current," labeled I_{co}.

Even with a reasonably large applied voltage, the current would be very small at room temperatures. This tells us that the resistance of the reverse-biased diode is very high. A typical value would lie in the range of $100,000\ \Omega$ to well over $1\ M\Omega$ (megohm).

2-3 FORWARD BIAS

We have seen that, except for a very small reverse current, there is no current flow through the diode if the N side is made positive and the P side is made negative by an applied voltage, notwithstanding the fact that each section by itself is a reasonably good conductor. We propose now to reverse the battery connections and investigate the difference, if any, between the two polarities.

We can see that the majority carriers would now be thrown *toward* the junction, rather than drawn away; this is depicted in Fig. 2-5. (Keep in mind the fact that the hole really does not exist as a separate particle. We say the holes are moving toward the junction; it is just as true that in order for a hole to move in one direction, an electron must move in the opposite direction.)

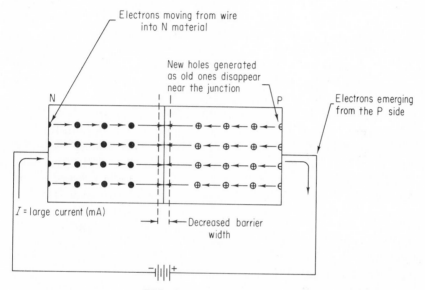

FIG. 2-5.   Forward bias.

As the holes and the electrons are moved toward each other, a large number of them will collide. As each hole is eliminated at the junction, a new one is formed somewhere in the volume of the P-type material. This must be true, because the number of holes is directly dependent upon the number of impurity atoms; and these, of course, cannot be changed or moved or destroyed. At all times there is a statistically constant number of holes present. As each hole is lost at the junction, an electron is just emerging at the wire connected to the P type. *The removal of an electron from the P side must result in the generation of a new hole.*

The same description could be applied to the N-type material. We cannot destroy the electron itself, only its effectiveness. So, for every electron lost in the process of recombination, a new one appears, supplied by the battery.

Now a large number of collisions are occurring at the junction, and it is seen that there is a large continuous current flow throughout the entire circuit. In view of this large current, we can deduce that the resistance of the diode is now *very low*. A typical value might be in the range of a fraction of an ohm to a few hundred ohms.

It might be instructive to trace the flow of current clear through the circuit from start to finish.

We shall follow a single electron around the complete circuit, starting at the negative post of the battery. Eventually, this electron will arrive at the terminal connected to the N-type material, and will be injected into the body of the crystal. It joins with the existing free electrons, and drifts to the right, impelled by the battery voltage, toward the junction. As it approaches the junction, a hole from the P side is moving toward it. Right at the junction, the two combine and are lost as separate entities. However, the hole, in order to have moved left toward our electron, must have been produced by a different electron that had moved to the right. Now *this* electron is the one that interests us, and we can see that we have simply "traded" electrons. Our new electron, then, begins a migration to the right, jumping from hole to hole, eventually emerging from the semiconductor to the right, then traveling along the wire to the positive post of the battery, through the battery, and back to our starting point.

We can make a few observations regarding this journey of an electron that may or may not be obvious.

1. The current at any point in the circuit is equal to the current at any other point.
2. If a given number of electrons are moving toward the junction in the N side, the same number of holes are moving toward the junction in the P side.
3. If a given number of electron-hole combinations are occurring per unit time at or near the junction, this number is equal to the number of electrons flowing per unit time past a point in the external circuit (wire).
4. The hole *cannot* exist outside of the semiconductor material. It must remain inside the confines of the diode itself.

In our discussion of diode forward bias, up to this point, we have neglected to consider the reverse current. With the diode forward-biased, we cannot properly apply the same name to this current, if it exists. In the N material the electron-hole pairs are still produced. The electrons thus generated join with the existing free electrons to become a part of the total available current carriers. In the P material, the thermally generated holes join with the existing hole to become a part of the existing current carriers. The normal forward current of a diode is, at normal temperatures, many thousands of times greater than this minute thermal current. So when a diode is forward-biased, the thermal current can usually be ignored.

2-4 DIODE SYMBOL

The symbol used in schematic drawings is shown in Fig. 2-6, along with the symbol we have used thus far. In either case, electron current

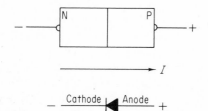

FIG. 2-6.   The diode symbol.

flows from left to right as shown. Also shown is the forward voltage necessary to cause this current flow.

2-5 DIODE CHARACTERISTICS

Diode characteristics are often shown in graphical fashion as in Fig. 2-7. The voltage labeled $+E$ is forward bias, and the voltage labeled $-E$ is reverse bias. By the same token, $+I$ is the current that flows when forward bias is applied, while $-I$ is the current that flows when reverse bias is applied.

In the forward direction current increases almost linearly for small increases in diode voltage ($+E$). In the reverse direction almost

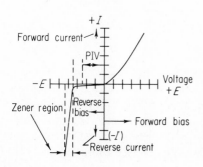

FIG. 2-7.   Diode characteristic curves.

no current flows until a certain reverse voltage is reached beyond the point labeled "PIV." The peak inverse voltage (PIV) is the maximum voltage that can be safely applied to a diode. Beyond this, the current again increases rapidly, and this region of operation is called the *zener region*. Operation in the zener region is destructive for the ordinary diode.

Certain diodes, however, are made especially to operate in this region, and they are called zener diodes. The symbol for such a device is shown in Fig. 2-8. The zener diode is widely used as a voltage reference, for the voltage drop across the diode remains essentially constant over a wide range of current values. Zener diodes will be discussed at length subsequently.

FIG. 2-8.   The zener symbol.

Consider the typical characteristic curve of a diode given in Fig. 2-9. This is quite similar to the one shown previously in Fig. 2-7, but with values for both current and voltage. At the point of origin, there is no current flow, of course, since there is no applied voltage. Hence, the curve passes through zero at this point. As the curve progresses to the right, it goes in the direction of increasing current and voltage. This suggests that if the current is increased in the diode, the voltage drop across it must also increase. At any point on the curve, the values of current and voltage that exist are represented. This is

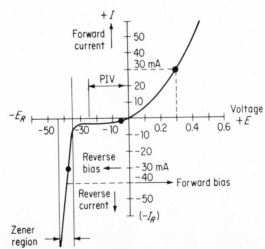

FIG. 2-9.   Forward and reverse diode characteristics.

the region of operation that is referred to as the *forward-biased region*.

Note especially that the diode is a current-dependent device; that is, the voltage drop across it is a result of allowing some value of current to flow through it. The current is limited to some particular value, and this causes the voltage drop to be some corresponding value consistent with the curve. Hence, to say that the voltage across a diode is going to be made equal to some value is not correct. Some amount of current is allowed to flow, and this will result in a value of voltage drop across the diode.

When the diode is reverse-biased, it operates in the region of the curve that is located to the left of the ordinate. As the voltage is increased, virtually no current flows, as shown by the curve. Even this small amount is exaggerated and is not drawn to scale. A typical value for this current is, perhaps, 1 to 10 μA (microamperes). However, as the voltage is increased beyond some critical value, a large current again flows. As the curve begins to slope downward very rapidly, this indicates a rapid increase in current. The diode is now operating in the avalanche, or zener, region, where, even though the diode is reverse-biased, current is again limited only by some external resistance in series with it. As mentioned before, a normal diode operating in the avalanche region is soon destroyed. To prevent this, the diode must be operated in the region below its peak-inverse voltage, which will prevent its destruction. The general description of the diode characteristic curve is seen to be rather simple.

To further explain the diode characteristic curve, assume that the simple circuit of Fig. 2-10 is to be used. If battery A (B_A) is connected to the terminals, the diode is reverse-biased, and the diode point of operation is indicated by a dot at −6 V (volts), just to the left of the point of origin in Fig. 2-9. This indicates that diode current is

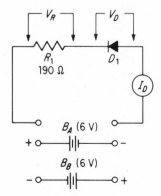

FIG. 2-10. A circuit to determine forward and reverse characteristics of a diode.

very small, consisting of only leakage current. At room temperature this can be assumed to be on the order of a few microamperes.

A typical manufacturer's data sheet specifies the maximum reverse current at rated voltage to be 1 μA for a particular silicon diode at 25°C (room temperature) and 10 μA at 100°C. Depending upon the diode in question, leakage current can range from a few nanoamperes to several hundred microamperes. Hence, the *relative amount* of reverse current as shown on the graph is grossly exaggerated.

Diodes that are used in certain applications, such as rectifiers or logic circuits, are always operated at reverse voltages less than their peak-inverse-voltage (PIV) rating to avoid the region where appreciable current again flows. As the reverse voltage reaches some critical value, the diode current begins to increase, and this can be attributed to one of two causes.

At low voltages, usually less than 10 V, the breakdown of the diode is due to *zener* action. Because the volume of material is very small, the voltage gradient within the diode material itself can reach values of several thousand volts per inch. The electrons are literally pulled from their covalent bonds and hence can contribute to current flow. The diode is then said to be operating in the zener region.

If the diode is designed to break down at higher voltages, the mechanism of current generation is slightly different. In this instance, the diode conducts because of *avalanche* breakdown. At higher voltages the first few carriers that break free attain much greater velocity than is the case in zener breakdown. These carriers, then, because of their greater velocity, actually collide with other bonded carriers, thus liberating them. These, in turn, liberate still others, and this cumulative action results in many carriers, which, of course, contributes to a heavy flow of current through the diode.

Such a device is properly called an *avalanche diode*, but it has become customary to identify all such diodes as *zener diodes*. While zener diodes function normally in this region of operation, normal diodes are quickly destroyed by such circuit action.

Now, when battery $B(B_B)$ is applied to the circuit, the diode becomes forward-biased. Note on the graph (Fig. 2-9) that the forward scale $(+E)$ is different than the reverse-voltage scale. The forward drop across the diode seldom exceeds 1 V, and so the scale is incremented in tenths of a volt. The diode-operating point is again indicated with a dot, and, as shown, the drop across the diode is about 0.3 V, with perhaps 30 mA (milliamperes) flowing. This can be proved simply by Ohm's law. The drop across the resistor is the supply voltage E_{BAT} less the diode drop V_D:

$$V_R = E_{BAT} - V_D = 6 - 0.3 = 5.7 \text{ V}$$

$$I_{R_1} = \frac{V_R}{R} = \frac{5.7}{190} = 30 \text{ mA}$$

The diode current, then, is indeed 30 mA.

Several further points concerning the circuit action of a diode should now be considered. First, the power dissipated by the diode is the product of V_D and I_D:

$$P_D = V_D I_D = 0.3 \times 0.03 = 0.009 \text{ W (watt)}$$

Note that the power dissipated by the diode is very small when forward-biased for moderate values of current.

To illustrate the damaging effect of operation in the reverse direction, assume that 30 mA of *reverse* current is to be allowed to flow. Now, the diode drop is about 40 V, and the power dissipation can be determined for this new condition:

$$P_D = V_D I_D = 40 \times 0.03 = 1.2 \text{ W}$$

Hence, operation in this region for any extended period of time will damage the diode unless it can easily dissipate, in this instance, 1.2 W. Note that while the current through the diode is the same in each instance, the dissipated power is much greater when reverse-biased. This is, of course, caused by the much greater voltage across the diode under these conditions.

Another factor that is often of interest is the static resistance offered to current flow. In the forward direction this resistance R_F is easily found:

$$R_F = \frac{V_D}{I_D} = \frac{0.3}{0.03} = 10 \ \Omega$$

Such a low value is, of course, to be expected.

The resistance offered to the circuit when the diode is reverse-biased but still within the PIV rating will be designated R_R and is a function of the applied voltage and the leakage current.

$$R_R = \frac{E_{BAT}}{I_{co}} = \frac{6}{1 \ \mu A} = 6 \ M\Omega$$

If leakage current is assumed to be 1 μA, the diode exhibits 6 MΩ of resistance. Compared with the usual circuit resistances, this is nearly an open circuit and is normally considered to be such.

Another value that is useful is the dynamic resistance of the diode when forward-biased. That is, the resistance offered by the diode to an alternating current while the diode is continually forward-biased is the ac, or dynamic, resistance. The symbol for this quantity as used herein is r_{df}. The value of r_{df} is determined by the physical and electrical conditions at the junction, as evidenced by the following equation:

$$r_{df} = \frac{KT}{QI_d m}$$

where T = absolute temperature, K (Kelvin)
  K = Boltzmann's constant [1.38×10^{-23} W-s/°C (watt-second/°Celsius)]
  Q = electron charge [1.6×10^{-19} C (coulomb)]
  m = constant (1 for Ge; between 1 and 2 for Si)
  I_d = diode current, dc

At room temperature this reduces to

$$r_{df} = \frac{0.026}{I_d}$$

As an example, a germanium diode having 1 mA of direct current has a dynamic resistance of $0.026/0.001 = 26$ Ω to an alternating current superimposed upon the direct current. If the direct current is increased to 10 mA, the ac resistance is reduced to $0.026/0.01 = 2.6$ Ω.

Another facet of diode operation is the static resistance offered by operation in the avalanche, or zener, region. Using the curve of Fig. 2-9, the resistance R_Z will be determined at 30 mA:

$$R_Z = \frac{V_D}{I_R} = \frac{40}{30 \text{ mA}} = 1333 \ \Omega$$

To illustrate the advantage of the zener diode used in regulating a voltage source, assume an increase in diode current to 40 mA. As read from the graph, the new voltage across the diode is on the order of 41 V, so a new value must be found:

$$R_Z' = \frac{V_D'}{I_D'} = \frac{41}{40 \text{ mA}} = 1025 \ \Omega$$

Note how drastically the diode resistance changes and how little the diode voltage changes. This is typical of zener action, and much larger variations of resistance are encountered in practice.

Because zener diodes are usually used in applications where the current is constantly changing, the dynamic resistance r_z is of greater importance than the static resistance. Using the two sets of numbers above will allow this value to be found:

$$r_z = \frac{\Delta V_D}{\Delta I_D} = \frac{V_D' - V_D}{I_D' - I_D} = \frac{1}{10 \text{ mA}} = 100 \ \Omega$$

Hence, the dynamic resistance of such a diode is much lower than the static value.

On the curve of Fig. 2-9 the slope of the diode characteristic in the zener region is exaggerated to clearly show that a slope exists. In an actual case, the slope would be much steeper (more nearly vertical), and thus the dynamic resistance would be much less. Typical values in the range of 0.1 to 100 Ω are found in current data sheets for low-power zener diodes.

Another interesting point concerning the general diode is that under certain conditions it acts as though it were a capacitor, the value of which is a function of the applied voltage. The barrier region of a diode acts very much like the dielectric of a capacitor since this region has essentially no current carriers available. Because the barrier width varies with the applied reverse voltage, the amount of capacitance exhibited also varies. Typical variation of capacitance is from less than 5 to perhaps 100 pF (picofarads). When diodes are designed to be used in this manner, they are called *varactors*, and they are often used in applications requiring variable capacitance that can be adjusted by changing the applied voltage. Many automatically operating circuits are based on this principle.

A simplified equivalent circuit, such as given in Fig. 2-11, is helpful in understanding the basic operation of a diode. In Fig. 2-11a the switch is open, which represents the reverse-biased junction. Note that the forward resistance R_{df} is effectively removed from the circuit. However, in parallel with both the junction and R_{df} is the reverse resistance R_r. Thus, some current will flow, but since R_r is typically very large in value, the reverse current is quite small. Should the frequency of the impressed voltage be reasonably high, then an alternate path exists through C_d. This, of course, represents the junction capacitance.

Figure 2-11b shows the forward-biased condition, and because the forward resistance is typically quite low (perhaps a few ohms to a

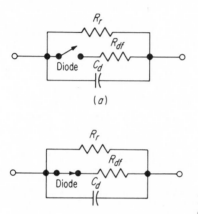

FIG. 2-11. Equivalent circuit of a diode: (a) the reverse-bias condition; (b) the forward-bias.

few hundred ohms), neither R_r nor C_d has any appreciable influence on diode action.

2-6 DIODE APPLICATIONS

Diodes are used to produce many different results. In this section, the usages will be illustrated and briefly described. It must be remembered that these examples are only representative of the dozens of ways in which diodes can be used.

Figure 2-12 illustrates four circuits that are commonly encountered. Figure 2-12a illustrates a negative-diode ground clamp. A circuit such as this is used to prevent an output excursion below the ground level. Regardless of the amplitude or polarity of the input voltage, the output can go no more negative than the normal forward-biased voltage drop across the diode. This is suggested by the waveforms, also shown. The circuit action, of course, is quite simple. When the input voltage is more positive than ground, the diode is reverse-biased and hence can in no way influence the output. When the input goes more negative than ground, the diode becomes forward-biased and in effect shorts out the incoming signal. The output is said to be clamped to ground during the negative half-cycle. The resistor is shown to emphasize the fact that there must be some current limitation to avoid damaging the diode.

Figure 2-12*b* shows a very similar circuit that differs only in the direction of the diode. Now, when the input goes more positive than

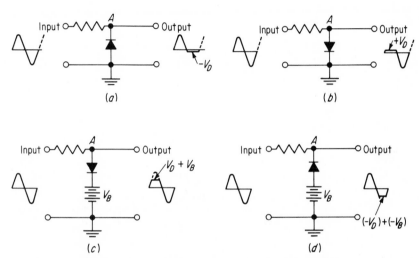

FIG. 2-12.   Four circuits illustrating diode applications.

ground, the diode becomes forward-biased and therefore clamps the output to ground during this half-cycle. Such a circuit is called a *positive-diode ground clamp.* The output voltage can go no more positive than the drop across the diode.

A similar circuit is shown in Fig. 2-12*c*, but it is different in that it employs a battery, which prevents the diode going into forward bias until a certain threshold voltage is exceeded. This circuit is called a *floating positive-diode clamp,* and its output can never exceed $V_B + V_D$. If the battery and diode are turned around, as shown in Fig. 2-12*d*, the output can go no more negative than $(-V_D) + (-V_B)$.

Diodes are also used as clippers, which are very similar to the clamp circuits. Figure 2-13*a* and *b* illustrates first a negative clipper and then a positive clipper. At the output in Fig. 2-13*a*, only when the input goes positive will there be an output since then the diode is reverse-biased. The negative spike is removed since the diode then becomes forward-biased, effectively shorting out the signal. With the diode reversed, Fig. 2-13*b*, the output spike is negative-going, with the positive-going part eliminated.

The circuits given in Fig. 2-13c and d are examples of diodes used as couplers. Again, the unidirectional characteristics of the diode are used to advantage. In the first case, the output can follow the input only when the input goes negative, while in the second case there is an output only when the input goes positive.

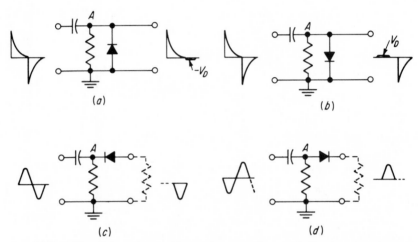

FIG. 2-13.   Four circuits illustrating coupling and clipping circuits.

The circuit shown in Fig. 2-14a is known as a *diode gate*, which is one of many so-called "conditional" circuits. A conditional circuit is one that produces an output only when the condition of the inputs agrees with a particular set of values. With both inputs firmly connected to ground (0 V), the output is also at a 0-V level. If either input A or B goes to a more positive voltage, the output *stays* at ground. Only when *both* inputs go to a more positive voltage will the output also go more positive. Such a circuit is often called an AND gate since both A AND B must be presented with a positive signal to produce an output. This kind of circuit will not be discussed in this book, since it occurs only in digital equipment.[1]

Figure 2-14b shows a diode rectifier. When an ac voltage is applied to the input, the output becomes pulsating direct current since the diode does not conduct during one half-cycle. Reversing the diode results in half-cycle output-voltage swings in the positive direction rather than in the negative direction.

[1] For a complete discussion of digital circuits, see Henry C. Veatch, "Pulse and Switching Circuit Action," McGraw-Hill Book Company, New York, 1971.

The circuits shown in Fig. 2-14c and d are widely used in digital applications. They serve to "steer" a pulse, or signal, toward a certain path, depending on the signal polarity. In Fig. 2-14c, if the input is more positive than ground, D_2 conducts and output 2 is positive. If, on the other hand, the input is more negative than ground, D_1 con-

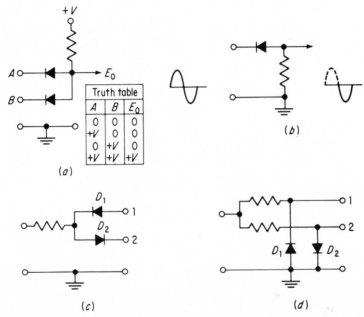

FIG. 2-14.   Gating, rectifying, and steering applications of diodes.

ducts and output 1 is more negative. Hence, positive and negative pulses are, in effect, separated and steered into their respective channels.

Figure 2-14d shows a slightly different arrangement, with the diodes in parallel with the signal. With an input more positive than ground, D_2 conducts and clamps line 2 to ground. However, with D_1 reverse-biased, line 1 is allowed to follow the input and thus goes positive. If the input goes in the negative direction, D conducts and line 2 is allowed to go negative, again channeling the pulses into their respective paths.

Diodes are often used for protection purposes, and an example of this is shown in Fig. 2-15a. With the switch closed, the inductor draws its normal current. The diode is at this time reverse-biased.

When the switch opens, the action of the inductor is to produce a large induced voltage (negative at the top, positive at the bottom), and the diode then becomes forward-biased, absorbing the excess energy in its very low internal resistance. The induced energy, then, is, in effect, removed from the surrounding circuitry, and the voltage across the coil will not exceed the normal forward drop across the diode.

A dc restorer circuit is shown in Fig. 2-15b. The circuit is first shown with no diode (and therefore no dc restoration) with attendant waveforms. However, with the diode connected in the circuit, the

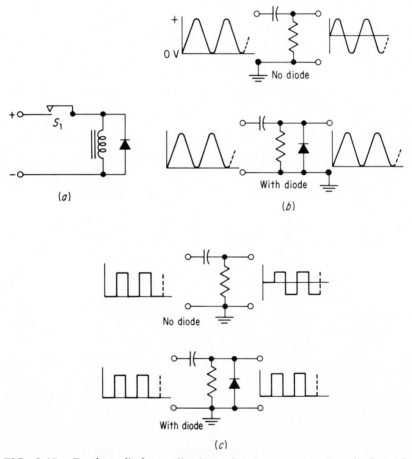

FIG. 2-15.   Further diode applications showing a protection diode and several circuits illustrating dc restoration.

output waveform can never go more negative than ground. If the diode were reversed, the output waveform would exist completely below 0 V, with the most positive value being ground. Figure 2-15c illustrates the dc restorer used with square waves.

The final example of diode circuitry is given in Fig. 2-16. With

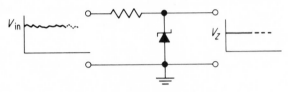

FIG. 2-16.   Zener diode application.

properly chosen values, the output voltage is constant in spite of input variations. Because the voltage across the zener diode is constant even though the current through it is varying, the voltage V_z is held constant. This circuit is called a *regulator* because of the smoothing or regulating effect of the diode on circuit voltage variations.

QUESTIONS AND PROBLEMS

True or false:

2-1 A PN junction is produced by the proper joining of P- and N-type silicon.

2-2 Pure germanium, as a semiconductor, exhibits a few free electrons and holes.

2-3 Forward bias reduces the area encompassed by the depletion region.

2-4 Reverse bias forces the majority carriers toward each other.

2-5 In an NP junction, when the P material is made more negative than the N material, this is called forward bias.

2-6 When a diode is reverse-biased, no current flows at all, under any circumstances.

2-7 Typical reverse-biased junction resistance is on the order of several hundred thousand ohms.

2-8 In a forward-biased diode, if the P-type material has twice the volume of the N-type material, there will be a heavier current flow on the P side of the junction than on the N side.

2-9 In a reverse-biased diode, the reverse current I_{co} is small at elevated temperatures.

2-10 A diode in forward bias has very low resistance.

2-11 When a diode is used as a rectifier, it is usually safe to exceed the PIV rating.

2-12 The proper use of diodes allows us to "steer" current in a pre-determined direction.

2-13 In the accompanying diagram, the source voltage is 12 V dc. Determine the total current ($V_D \cong 0$ V).

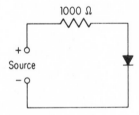

2-14 Refer to the diagram of Question 2-13. The source voltage is 12 V, and the drop across the series resistor is 12 V. Determine (a) the current in the circuit and (b) the dc resistance of the circuit.

2-15 Refer to Question 2-14. A new diode is to be placed in the circuit, and V_D when measured is 0.3 V. Determine (a) the circuit current and (b) the diode dc resistance.

2-16 Refer to Question 2-14. A new diode is to be installed, and its V_D is known to be 0.7 V. Determine (a) the circuit current and (b) the dc diode resistance.

2-17 A diode is placed in a circuit, and the diode voltage drop is found to be equal to the supply voltage of 10 V. The diode current is measured as 10 μA. Find the value of R_R.

2-18 A diode is placed in a circuit, and the diode voltage drop is found to be equal to the supply voltage of 10 V. The diode current is measured as 25 μA. Find the value of R_R.

2-19 A germanium diode is operated at a direct current of 3 mA. Determine the dynamic resistance offered to an ac signal.

2-20 A germanium diode is operated at a direct current of 0.5 mA. Determine the dynamic resistance offered to an ac signal.

2-21 A zener diode has a nominal zener voltage of 20 V. The direct current is 30 mA. Find R_Z.

2-22 A zener diode has a nominal zener voltage of 20 V. The direct current is 40 mA. Find R_Z.

2-23 Refer to Questions 2-21 and 2-22. The two separate conditions refer to the same diode. Find r_z if the voltage across the diode increases 0.1 V as the current increases by 10 mA.

2-24 Refer to Questions 2-21 and 2-22. The two separate conditions refer to the same diode. Find r_z if the voltage across the diode increases 0.01 V as the current increases by 10 mA.

CHAPTER

3 THE JUNCTION TRANSISTOR

3-1 THE TWO-JUNCTION DEVICE

The semiconductor diode discussed in Chap. 2 is closely related to a transistor. In some respects, their actions are very similar. The main difference is that the transistor has *two* junctions rather than one.

This means that there must be three sections of material instead of two. Such a device is shown in Fig. 3-1. We shall assign a name to each of the three parts that will be descriptive of the function that each is to have in the complete transistor. The left-hand part of N-

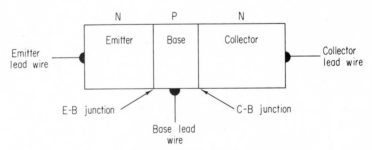

FIG. 3-1.   NPN transistor.

type material is labeled the "emitter," and we shall expect that its function is to emit something. The right-hand part, also of N-type material, is called the "collector," and again the purpose will be to collect something. The intermediate area is common to both ends, and so we say that it is the "base." Such a configuration is known as an NPN type.

It is possible to connect externally generated voltages to the transistor, and to determine how each part works by measuring the resultant currents. From these currents, we can discover how the transistor is capable of performing all the various jobs that it does so well. Consideration of Fig. 3-1 will reveal that there are four possible ways of connecting the transistor to external voltages. We can enumerate these in outline form.

1.   Emitter + with respect to base
    a.   Collector + with respect to base
    b.   Collector − with respect to base
2.   Emitter − with respect to base
    a.   Collector + with respect to base
    b.   Collector − with respect to base

Of these four possible combinations, only one interests us at the moment: the condition where the emitter is negative with respect to the base and the collector is positive with respect to the base. So that we can visualize the occurrences inside the transistor, we shall modify the drawing of Fig. 3-1 slightly, as in Fig. 3-2. Note that the changes are the addition of the three current meters, two batteries, and two switches. This circuit arrangement will allow us to apply voltage when

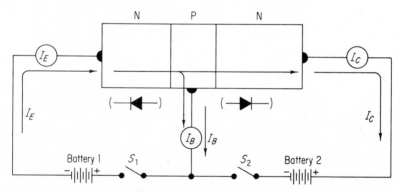

FIG. 3-2.   Transistor biases.

necessary and to measure the resultant current. The collector current is called I_C, the emitter current is I_E, and the base current is I_B.

If only switch S_1 is closed, we shall be applying the battery voltage to the emitter and the base, causing the emitter to be more negative than the base. The symbol just below the emitter-base junction indicates an equivalent diode, the implication being that the emitter-base junction will behave very much like the diode symbol. Applying what was learned in Chap. 2, we should be forced to conclude that the emitter-base junction is forward-biased.

If this is true, and it is, a relatively large current is flowing in the circuit, starting at the battery, up through the meter I_E, to the emitter, across the emitter-base junction, through the base, and out to the external circuit through the meter I_B, and back to the battery. We should note equal currents flowing in the base lead and the emitter lead, as is to be expected.

When S_1 is opened and S_2 closed, we should note that the collector-base junction is reverse-biased, and *no* current would be indicated at either the I_C or I_B meters.

Each of the above circuit conditions is perfectly straightforward, and is exactly what we should expect. Each of the junctions is performing just like any junction diode, so long as only one switch is closed at a time.

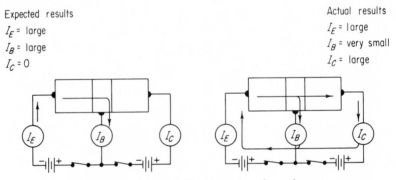

FIG. 3-3.   Expected versus actual results.

We have so far studiously avoided closing both switches together, but now we must ask: What should we expect to occur if both S_1 and S_2 are closed? Well, it would seem that meters I_B and I_E would yield large readings and I_C would yield a zero reading since simultaneous readings should not be much different from individual readings. Figure 3-3 shows the expected versus the actual results of simultaneous switch closure, and we see that the base current becomes *very small* and the collector current *very large*.

This result is totally unexpected, and the reason for I_C being large, while I_B is small, must be investigated.

Figure 3-4 depicts the three sections of a typical NPN transistor. Only the majority carriers are shown, for simplicity. Also shown are the barrier regions, one at the emitter-base junction and one at the collector-base junction. Because of the forward bias at the emitter junction, this barrier is very small in width. At the collector junction the barrier width is increased, since the reverse bias pulls the majority carriers away from the junction.

Normally, there are no carriers present in the barrier region, but when the transistor is made to operate, some of the electrons going from the emitter into the base find their way to the collector junction. Once at the junction, the collector supply voltage can be felt by the electrons. They now proceed around the circuit to become the normal collector current. One reason why the electrons move across the base without finding a hole with which to recombine is that the base region is deliberately made very thin during manufacture. Because of this there are few majority carriers (holes) in

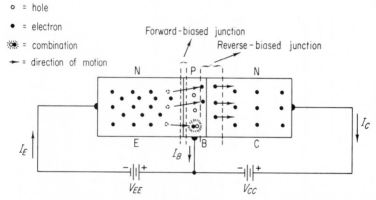

FIG. 3-4.   Transistor amplifier biases, NPN.

the base region. Thus the electrons from the emitter have difficulty finding a hole to combine with. In addition, the base region is doped very lightly, and this is another reason for fewer holes in the base region.

The light doping and the thinness of the base [0.001 in.(inch)] contribute to the few recombinations in the base region. Generally, the fewer recombinations the better; thus the smaller the base current the better.

Still referring to Fig. 3-4, let us visualize the occurrences in a conducting transistor. With voltage applied, the emitter injects its majority carriers into the base region because of its forward bias. Many electrons are injected, many more than there are holes to recombine with. The electrons that cannot find a hole to combine with are forced toward the collector junction by the new electrons arriving from the emitter, and they will diffuse across the junction. Once across the junction, they are attracted by the large V_{CC} and become a part of the collector current. The number of electrons that cross the emitter junction is determined by the reduced barrier width of the emitter-base junction, which in turn is determined by the magnitude of forward bias. Thus, increasing the forward bias causes an increase in emitter current, with the corresponding increase in both base and collector current. Likewise, a decrease in forward bias results in a decrease in all current.

In order to better understand the mechanism of current flow within the transistor, let us trace current flow throughout the entire circuit. Starting at $-V_{EE}$, the electrons flow out and into the emitter lead. These electrons move into the emitter material in normal fashion up to the base junction. Depending upon the amount of forward bias, some number will be allowed to pass by the junction and enter the base region. A few of these electrons will find holes and will recombine, producing a small base current. Most of them, about 98 percent, will not find a hole because there are so few of these. So they are literally pushed to the collector junction by those coming from behind. Because they are constantly agitated by thermal activity, they wander across the junction. Once across the collector junction, they feel the influence of V_{CC} and are swept toward the collector lead. They then return to V_{CC}, pass through the battery, and join with a small number of electrons coming from the base. They enter V_{EE} and return to their starting point at $-V_{EE}$.

The foregoing discussion pertained to an NPN-type transistor. Equally important is a PNP type, and we shall note that there are only two major differences in these two types. The majority carriers are different in corresponding parts of the transistor. That is, in the NPN type the carrier is the electron in the emitter, while in the PNP type the majority carrier is the hole. Also, the applied battery voltages are the opposite in each case.

Only a brief description of the PNP-type transistor will be given because of the similarity of the two types. Taking into account the interchange of majority carriers between the two types and the directions that the external currents will flow, we can see in Fig. 3-5 that the majority carriers, that is, holes, are injected into the base region. Again, the base region is lightly doped and very thin, and there are very few electrons with which the holes can combine. Most of them therefore diffuse across the collector junction and become collector current. While in the base region, a very few of them will recombine and will contribute to a very small base current. Most of them, however, continue on to become a part of the collector current. As the holes move from left to right on the drawing, electrons must, of course, be flowing from right to left. The electron flow is from $-V_{CC}$ toward the collector, through the collector, through the base, through the emitter, and out the emitter wire into the external circuit.

Note that, as was the case with a diode, the holes cannot leave the

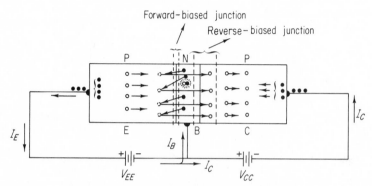

FIG. 3-5. Transistor amplifier biases, PNP.

semiconductor material. As before, the holes that reach the physical limits of the P-type collector material are annihilated by entering electrons and cease to exist as current carriers. But, at the same time, electrons are leaving the P-type emitter material, and as a hole is annihilated at the collector terminal, a new one is created at the emitter terminal.

Therefore, there is a constant number of holes existing at all times, the number of them being dependent only upon the number of impurity atoms originally introduced during manufacture.

Both types, then, are seen to be quite similar, the only differences, as noted, being in the direction of external current flow and in the interchange of holes and electrons in the two types.

3-2 THE TRANSISTOR AS A CIRCUIT ELEMENT

Symbols

Before we can begin to understand how a transistor functions in a typical circuit, we shall find it necessary to use symbols for our drawings that will properly represent the transistor and will be simple to execute. The standard symbols are shown in Fig. 3-6.

The only difference between the two symbols is in the direction of the arrowhead on the emitter, and this simply says that current will flow in opposite directions in the two types. Electron current is *against* the arrowhead. In order to show the method of connection and the proper supply voltages, we shall connect each into an appropriate circuit, very much as we did in Figs. 3-4 and 3-5. This is illustrated in Fig. 3-7, where the various currents are shown for both PNP and

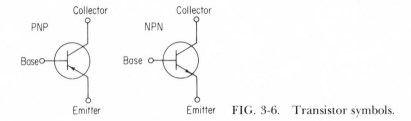

FIG. 3-6.   Transistor symbols.

NPN types. Note that all transistor current flows in the emitter lead. The emitter current, then, must be the sum of the collector current and base current. This can be written as a formula:

$$I_E = I_C + I_B$$

This relationship always holds true. If a value is given for any two of the currents, the third can be found by one of the following expressions:

$$I_E = I_C + I_B$$
$$I_C = I_E - I_B$$
$$I_B = I_E - I_C$$

In Fig. 3-7, if we use either of the drawings as an example to determine the relative magnitude of currents, we can see that the place in the circuit where the maximum current is flowing is in the emitter lead. Instead of assigning some number of milliamperes to

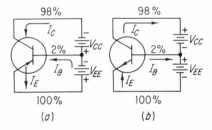

FIG. 3-7.   Division of current. (*a*) PNP. (*b*) NPN.

I_E, let us talk about this current in terms of percent of maximum. I_E then must be 100 percent, since all current that is flowing anywhere in the circuit must pass through this point.

We should then find that the base current I_B would be about 2 percent of the total current. If this is true, this means that $100 - 2 =$ 98 percent of the emitter current is flowing in the collector lead. The base current, we recall, is the result of the recombination occurring in the area of the base, and the thinner the base, the smaller the base current.

We can see that in order to have any collector current at all, there must be carriers injected from the emitter into the base region. This is accomplished by forward-biasing the emitter-base junction. If at any time the forward bias is removed, there will be no I_C. This simply means that if we want to "turn off" the transistor, we need only remove the forward bias from the emitter-base junction.

A transistor is said to be "on" if V_{CC} is applied and if the emitter-base junction is forward-biased. Under these conditions, collector current will flow and the transistor is on. If the emitter-base junction is not forward-biased enough to cause some base current, and therefore some emitter current, then no collector current will flow, and the transistor is off.

Leg-Current Relationships

The percentage values for the three transistor currents will be found to vary widely from transistor to transistor. In any single transistor, however, the percentage of collector current, compared with the emitter current, is reasonably constant. That is, if a transistor has its currents measured, and if collector current is 98 percent of the emitter current, this transistor will always have a collector current that is about 98 percent of its emitter current. There are several factors that will cause this percentage division to vary, including temperature, current density, frequency, and voltage. The deviation is usually slight for most practical cases, and it is useful to think of this current division as constant.

Alpha

Because the relationship described above is considered a constant, it is given a name, the Greek letter alpha (α). The alpha of a transistor is simply the ratio of collector current to emitter current, and it is always less than 1. For example, if a transistor has an emitter current of 21 mA, and if 20 mA of collector current is flowing, the alpha is computed as follows:

$$\alpha = \frac{I_C}{I_E} = \frac{20 \text{ mA}}{21 \text{ mA}} = 0.952$$

If, in this same transistor, the emitter current is increased to 31.4 mA, the collector current will increase to about 30 mA. Alpha is still essentially the same.

$$\alpha = \frac{I_C}{I_E} = \frac{30 \text{ mA}}{31.4 \text{ mA}} = 0.955$$

However, if the current were to be made many times larger, say, 50 times, then the manufacturer's data sheets would have to be referred to in order to determine how the alpha changes with so large an increase.

Beta

There is one more transistor constant that is even more useful. By ratioing collector current to base current, we determine the beta (β) of the transistor. *The constant alpha tells us how much less I_C is than I_E. Beta tells us how much greater I_C is than I_B.*

In the foregoing example ($I_B = I_E - I_C = 21 \text{ mA} - 20 \text{ mA} = 1 \text{ mA}$), the transistor beta is

$$\beta = \frac{I_C}{I_B} = \frac{20 \text{ mA}}{21 - 20 \text{ mA}} = \frac{20 \text{ mA}}{1 \text{ mA}} = 20$$

Thus, for this transistor, beta is 20, and this tells us that the collector current is 20 times the base current. As a simple example of how these factors are used, consider the following discussion.

If the emitter current in a given situation is doubled, both collector current and base current will double. Suppose a transistor has, originally, the following currents flowing: I_E is 1 mA, I_C is 0.98 mA, and I_B is 0.02 mA. If any one of these is caused to change by 2 times, the others must also change by the same factor. Thus, if I_B is increased to 0.04 mA, I_E must become 2 mA (2×1 mA), and I_C will become 1.96 mA (2×0.98).

In the foregoing example, note that we caused the base current to change from 0.02 to 0.04 mA, a 0.02-mA change. This resulted in a

0.98-mA change in the collector circuit current. This can rightfully be called *current amplification*, since a large change was produced by a small change. We shall subsequently have more to say about current amplification.

Figure 3-8 shows pictorially how the currents respond to changes.

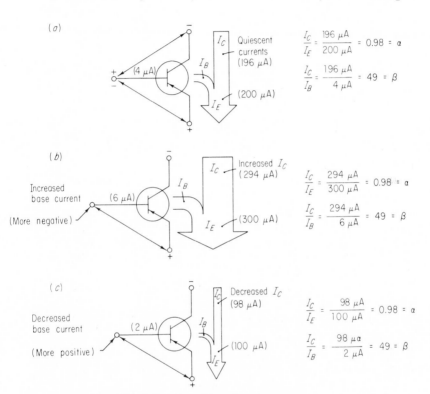

FIG. 3-8. Amplifier signal currents.

Figure 3-8a shows the original amount of current flowing before any change. These values are called *quiescent* values of current. For this example I_B is 4 μA, I_C is 196 μA, and I_E is 200 μA.

If the base current is caused to increase to 6 μA, both collector and emitter current must increase also. But the question is: How much will I_C and I_E increase?

From Fig. 3-8a we can determine the beta of the transistor:

$$\beta = \frac{I_C}{I_B} = \frac{196 \ \mu A}{4 \ \mu A} = 49$$

This says that I_C is always 49 times greater than I_B. If I_B is increased by 2 μA, as shown in Fig. 3-8b, then I_C *must increase beta times the base-current increase.*

$$I_C = \beta I_B \text{ increase} = 49 \times 2 \ \mu A = 98 \ \mu A \text{ increase}$$

If, quiescently, I_C is 196 μA, and is caused to increase by 98 μA, it is now $196 + 98 = 294 \ \mu$A, as shown.

By knowing that $I_C = \beta I_B$, we should also calculate the new I_C by the following relation:

$$I_C \text{ increase} = \beta I_B \text{ increase} = 49 \times 6 \ \mu A = 294 \ \mu A$$

The emitter current increases by the factor $(\beta + 1)$; thus the new value is $(\beta + 1) \times I_B = (49 + 1) \times 6 \ \mu A = 50 \times 6 \ \mu A = 300 \ \mu A$. If base current is caused to decrease, collector and emitter currents must also decrease. (See Fig. 3-8c.)

The new values are

$$I_C = \beta I_B = 49 \times 2 \ \mu A = 98 \ \mu A$$
$$I_E = (\beta + 1) I_B = 50 \times 2 \ \mu A = 100 \ \mu A$$

Note that for all three conditions the percentage relationships between all three currents still hold true.

During the time that currents are caused to change, we say that the transistor is operating under *signal conditions.* That is, just as the steady-state, or dc, conditions are called the quiescent conditions, the ac conditions are signal conditions.

A transistor circuit requires both conditions to perform useful work. The dc conditions are often called the bias conditions, and are necessary before a signal is applied. Once a transistor is biased properly, a signal current can be applied, and the transistor will then amplify. In the example of Fig. 3-8 our input is the signal current applied to the base, and its peak-to-peak value is $6 - 2 = 4 \ \mu$A. The output current I_C has a peak-to-peak value of $294 - 98 = 196 \ \mu$A.

Since the input signal is 4 μA peak to peak and the output signal is 196 μA peak to peak, the transistor is amplifying. The amount of amplification β_{ac} is determined by dividing the output-current change by the input-current change:

$$\beta_{ac} = \frac{I_{out}}{I_{in}} = \frac{\Delta I_C}{\Delta I_B} = \frac{196 \ \mu A}{4 \mu A} = 49$$

We recognize this number as being beta for this transistor, as it should be.

To further describe the amplifying properties of a transistor, we show Fig. 3-9. This is the same example as Fig. 3-8, but shown a different way. The same description is valid, and therefore it will not be repeated here.

Because both alpha and beta are derived from the percentage division of current, it would seem that they are related. If we are

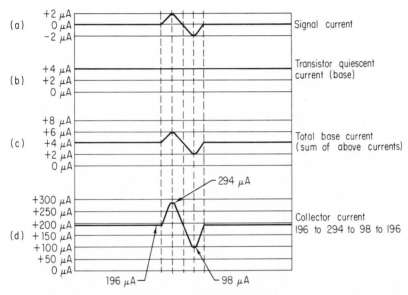

FIG. 3-9.   Signal waveforms.

given one, we can easily determine the other by the following relationships:

$$\alpha = \frac{\beta}{1 + \beta} \qquad \beta = \frac{\alpha}{1 - \alpha}$$

Using the same numbers as before, we shall work this out to see if it is true.

$$\alpha = \frac{196}{200} = 0.98$$

$$\beta = \frac{0.98}{1 - 0.98} = \frac{0.98}{0.02} = 49$$

and we arrive at the same β as before.

A summary of the formulas in this section is given below. All are useful in analyzing transistor circuits.

$$I_C = I_E - I_B = \beta I_B$$
$$I_E = I_C + I_B = (\beta + 1)I_B$$

$$I_B = I_E - I_C = \frac{I_C}{\beta} = \frac{I_E}{\beta + 1}$$

$$\alpha = \frac{\beta}{1 + \beta} = \frac{I_C}{I_E}$$

$$\beta = \frac{\alpha}{1 - \alpha} = \frac{I_C}{I_B}$$

3-3 TRANSISTOR EQUIVALENT CIRCUIT

To assist in visualizing the action of a transistor in a circuit, an equivalent circuit is often used. Many different equivalent circuits have been devised, from the very simple to the very complex. Generally, the

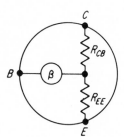

FIG. 3-10. Bipolar transistor equivalent circuit.

simplest one that allows the computation of the desired results is the best one to use.

A very simple equivalent circuit is shown in Fig. 3-10. Because the three sections of the transistor are conductive to a greater or lesser degree, the collector and emitter regions can be shown as resistances. The generator symbol shown in the base lead indicates the current amplification that exists. This has a numerical value equal to the beta of the transistor.

Figure 3-11 shows the static condition for a typical NPN transistor in a simple circuit. The current generator (beta) for this example is equal to 50. Hence, the resistor RB has been chosen to allow 20 μA (0.02 mA) to flow in the base circuit. Collector current is therefore β times I_B:

$$I_C = \beta I_B = 50 \times 0.02 = 1 \text{ mA}$$

Emitter current is $\beta + 1$ times I_B:

$$I_E = (\beta + 1)I_B = 51 \times 0.02 = 1.02 \text{ mA}$$

To allow a complete description of the circuit action, values for the collector resistance (R_{CB}) and emitter resistance (R_{EE}) of the transistor itself must be determined. As will be discussed in detail in a later chapter, the value of R_{EE} can be determined from known transistor characteristics, while the value of R_{CB} is actually a function of applied voltages and currents.

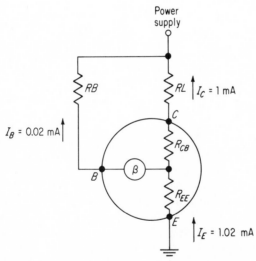

FIG. 3-11. Static conditions of simple transistor circuit.

QUESTIONS AND PROBLEMS

Select the correct answer for each of the questions below.

3-1 Hole current is the movement of
    (*a*) positive charges in the opposite direction from electron current
    (*b*) positive charges in the same direction as electron current

(c) negative charges in the opposite direction from electron current

(d) neutral charges in the opposite direction from electron current

3-2 The barrier voltage at a PN or NP junction (germanium) is approximately
(a) 0.2 V   (b) 2 V
(c) 25 V   (d) 2 mV

3-3 The collector junction in a transistor amplifier circuit has
(a) reverse bias for NPN and forward bias for PNP
(b) reverse bias for PNP and forward bias for NPN
(c) reverse bias at all times
(d) forward bias at all times

3-4 The emitter junction in a linear transistor amplifier circuit has
(a) reverse bias for NPN and forward bias for PNP
(b) reverse bias for PNP and forward bias for NPN
(c) reverse bias at all times
(d) forward bias at all times

3-5 The battery connections required for proper operation of a PNP amplifier circuit are
(a) + to collector and − to emitter; − to base and + to emitter
(b) − to collector and + to emitter; + to base and − to emitter
(c) + to collector and − to emitter; + to base and − to emitter
(d) − to collector and + to emitter; − to base and + to emitter

3-6 The arrow on the symbol for a PNP transistor is drawn in the direction of
(a) electron current in the emitter
(b) hole current in the collector
(c) hole current in the emitter
(d) electron current in the collector

3-7 A transistor is made to increase its base current from 30 to 40 μA. It is noted that the collector current increases from 0.5 to 0.9 mA. The β_{ac} of this transistor is
(a) 0.975   (b) 0.944
(c) 40   (d) 16.7

3-8 A certain transistor is described as having a beta of 40. What is the alpha?
(*a*) 41   (*b*) 0.932
(*c*) 0.976   (*d*) 0.998

3-9 Refer to Fig. 3-7. In either transistor circuit, $I_C = 95$ percent, $I_B = 5$ percent, and $I_E = 100$ percent. Determine the values of alpha and beta.

3-10 Refer to Fig. 3-7. In either transistor circuit, $I_C = 99$ percent. Find the percentage of I_B.

3-11 Refer to Fig. 3-7. In either transistor circuit, $I_C = 97$ percent, $I_B = 3$ percent, and $I_E = 100$ percent. Base current is known to be 0.01 mA. Determine the value of I_C.

3-12 Refer to Fig. 3-7. In either transistor circuit, if the transistor's alpha is 0.975, find the relative percentage of I_C, I_B, and I_E.

3-13 A given transistor has a base current of 0.02 mA when emitter current is 2.02 mA, and a base current of 0.03 mA when emitter current is 2.9 mA. At which value of emitter current is beta the highest?

CHAPTER 4 CHARACTERISTIC CURVES

4-1 GENERAL DESCRIPTION

If we were to ask the manufacturer of a transistor to supply us with data describing the various ways the unit could be used, and if all the data were supplied in tabular form, it would take us hours to pore over the information to find the one point of interest to us. So many different combinations of voltages and currents are possible with a given unit that a different scheme had to be devised to describe, in technical terms, just how a transistor will work under a variety of conditions. The most compact way of describing a transistor is by

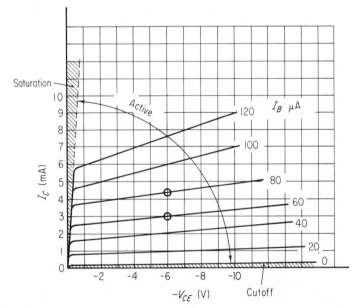

FIG. 4-1. Output characteristic curves — regions of operation.

plotting on a graph the necessary information. We call such a graph the *characteristic curves* of a transistor.

Such a set of curves, the collector curves, is shown in Fig. 4-1. One ordinate of the graph represents collector current I_C, while the other is collector-to-emitter voltage V_{CE}. The running parameter (slanted lines) represents base current I_B. From these curves we can read the various currents and voltages that must exist for different operating conditions.

4-2 READING THE COLLECTOR CHARACTERISTIC CURVES

To take an example, there is a point on the curves that is encircled on the 60-μA base-current line. If the transistor that these curves describe is operated with 60 μA of base current, and if V_{CE} is 6 V, 3 mA of collector current is flowing. This is read by projecting a line straight left from the intersection of an $I_B = 60$ μA and a V_{CE} of 6 V. Any combination of I_C, I_B, and V_{CE} that the transistor can attain is to be found on the curves.

Knowing that $I_E = I_C + I_B$, we can also determine emitter current for any point on the curves. In the preceding example, emitter current is 3 mA + 60 μA, or 3.06 mA.

We can use the curves to derive characteristics of the transistor other than the currents and voltages. Since beta is I_C/I_B, it is possible to derive this important parameter by using the curves. Taking the same example as before,

$$\beta = \frac{3 \text{ mA}}{60 \text{ } \mu\text{A}} = \frac{0.003}{0.00006} = 50$$

By the same token, alpha is I_C/I_E. Therefore

$$\alpha = \frac{0.003}{0.00306} = 0.9804 \cong 0.98$$

Because they were derived from dc values, they are dc values of alpha and beta. Direct-current beta is often written h_{FE}.

4-3 DELTA VALUES

It now becomes necessary to modify slightly the relationship described above. I_C/I_B is called the dc beta (h_{FE}), since the values of I_C

and I_B used in this expression are static, unchanging values. The so-called ac beta, or h_{fe}, is much more useful since we normally apply an ac signal to the transistor to be amplified. This implies a changing input with a resulting changing output. We can simulate an ac signal applied to the base of a transistor without even so much as seeing the unit itself if we have its characteristic curves at hand. Suppose the input signal is such as to cause a change of base current from 60 to 80 μA, with the voltage between collector and emitter equal to 6 V. If we locate these two points on the curve of Fig. 4-1, we can see that the base current made a change of $80 - 60 = 20$ μA, which is the difference between the two values. We write this difference as $\Delta I_B = 20$ μA, which is read as "delta I_B." The delta sign means simply that the value indicated is a *difference* between two separate values without regard to the two values themselves. For instance, $\Delta I_B = 20$ might be derived from $80 - 60$ as above, or $100 - 80$, or $30 - 10$, or any other set of numbers whose difference is 20. This provides us with a convenient and useful way of denoting such differences without having to refer to the numbers from which the difference is derived.

$$\text{If } \beta_{dc} = \frac{I_C}{I_B} \quad \text{then } \beta_{ac} = \frac{\Delta I_C}{\Delta I_B}$$

The junction of the 60-μA base-current line and the V_{CE} 6-V line is encircled in Fig. 4-1, and from this point straight left is the amount of collector current that is flowing at this instant. We read from the ordinate, $I_C = 3$ mA. Now, if 80 μA is caused to flow in the base, the collector current will increase to 4.25 mA. Again, read from the ordinate, opposite the 80-μA and 6-V junction, which is encircled, 4.25 mA. (V_{CE} stays the same.) Now compiling these four numbers, we can say: If

$$I_B = 60 \ \mu A \qquad \text{then } I_C = 3 \text{ mA}$$

and if

$$I_B = 80 \ \mu A \qquad \text{then } I_C = 4.25 \text{ mA}$$

We can now determine the β_{ac} or h_{fe}, for this transistor:

$$\beta_{ac} = h_{fe} = \frac{\Delta I_C}{\Delta I_B} = \frac{(4.25 \times 10^{-3}) - (3 \times 10^{-3})}{(80 \times 10^{-6}) - (60 \times 10^{-6})} = \frac{1.25 \times 10^{-3}}{20 \times 10^{-6}}$$
$$= 0.0625 \times 10^3 \cong 63$$

and so the ac amplification is approximately 63.

To verify that β_{dc} is often different from β_{ac}, use any one set of the above numbers such as

$$\beta_{dc} = h_{FE} = \frac{3 \times 10^{-3}}{60 \times 10^{-6}} = \frac{3 \times 10^{-3}}{6 \times 10^{-5}} = 0.5 \times 10^2 = 50$$

which is indeed a different value of β.

What we have done above is to use the curves to derive some of the characteristics of the transistor.

4-4 REGIONS OF OPERATION

By referring again to Fig. 4-1, we can use these curves to describe the three possible regions of operation. A transistor operating in the center portion of the curve is said to be in the *active* region. This is one of the three possible regions of operation. A transistor can amplify only when it operates in the active region.

A second possible region is called *cutoff*. On the curves, this is the area near the abscissa, where $I_C = I_{cbo} \cong 0$. I_{cbo} is the transistor leakage current, being very similar to I_{co} in a diode. When a transistor is in cutoff, no collector current or base current flows, except for leakage current, and the transistor acts very much like an open switch.

The third region of operation is called *saturation*. This is the area of operation where maximum collector current flows, and the transistor acts like a closed switch. The transistor is said to be *full on*.

When a transistor is in the saturation region or in cutoff, not only can it not amplify, but the normal relationships (α, β) between the three leg currents do not hold true.

To see how these regions of operation are applied to the principles of circuits, consider Fig. 4-2. The first drawing depicts operation in the active region, where V_C is something less than V_{CC}, but not quite ground. Here I_C is β times I_B, and I_E is ($\beta + 1$) times I_B, and the transistor is said to be biased in the active region.

Figure 4-2b depicts the cutoff region, where I_C, I_E, and I_B are essentially zero. No current through the 1-kΩ (kilohm) load resistor results in no drop across it; so V_C is equal to $-V_{CC}$. Obviously, the normal relationships between the currents cannot exist since there is essentially no current.

Finally, the saturation region is shown, where V_C is nearly at ground, and collector current is limited solely by V_{CC} and RL.

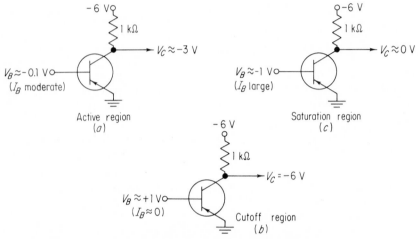

FIG. 4-2.   Three regions of operation.

OTHER TRANSISTOR CURVES

The characteristic curves previously described, called the collector output curves, are extremely useful for determining how a transistor functions in terms of its voltages and currents. Other sets of curves reveal still more about how a transistor will work under a given set of conditions. For example, the so-called input curves tell us how the base-to-emitter voltage V_{BE} changes with changes in collector current. Generally, as collector current is increased, and therefore I_B is increased, the base-emitter voltage will increase. But this increase is not necessarily linear, and so is best shown as a set of curves. Because V_{BE} depends, for one thing, upon temperature, the curves are often related to the junction at several different temperatures. As another example, the manufacturer often supplies curves describing how the beta of a transistor changes with temperature. As the temperature increases, beta increases, but by a different amount for different values of collector current. Again, this change is not a linear one, and so is best presented as a curve. These curves, as well as others, are usually of interest to the engineer, but at this time we merely note their existence. As we go further into the subject of transistors, we become more aware of the significance of this additional information, and then are better able to digest it. We shall later investigate some of these areas of interest. The general subject of transistor circuits is difficult, at best, and the more information we have at hand, the better able we are to do a job. In this chapter we shall only introduce these curves. Section 4-5 describes them briefly.

4-5 INPUT AND TEMPERATURE CURVES

In addition to the V_{CE}–I_C curves, the manufacturer of transistors will provide other sets of curves to describe the various characteristics of the product. One of these, often supplied, relates to information about the input characteristics of the transistor.

On these curves, collector current I_C is plotted versus base-emitter voltage V_{BE}, often for different temperatures. Such a curve is shown in Fig. 4-3 for a typical germanium transistor. A similar set is shown in Fig. 4-4 for a silicon NPN transistor. Note the greater V_{BE} for the silicon unit, which is characteristic for this type of transistor.

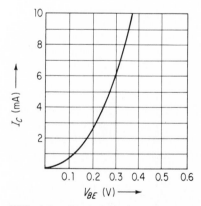

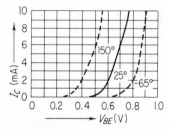

FIG. 4-3.   Transistor input curves, Ge.

FIG. 4-4.   Transistor input curves, Si.

These input curves describe how the base-emitter voltage V_{BE} varies as the collector current I_C is varied. Any given point along the curve gives the value for these conditions of the ohmic emitter resistance R_{EE}.

Using the curves for the germanium transistor as an example, if

V_{BE} is 0.2 V and I_C is 9 mA, R_{EE} is $0.2/0.009 = 22\ \Omega$. If the curves are given, it is a simple thing to determine the value of R_{EE}.

Another set of curves describes how the transistor parameters vary with the temperature. One way of showing this is illustrated in Fig. 4-5. Here the variation of beta with temperature is shown. As the temperature increases, beta (h_{FE}) increases. At about 100°C its value has about doubled. This, of course, can cause problems with the temperature stability of a transistor circuit.

Figure 4-6 illustrates how leakage current I_{cbo} varies with temperature. In Chap. 11 we shall discuss this curve in some detail.

The final curve that we shall show here, in Fig. 4-7, is a typical

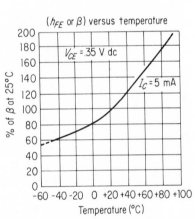

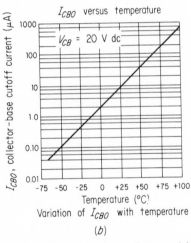

FIG. 4-5. Variation of beta with temperature.

FIG. 4-6. Variation of I_{cbo} with temperature. (*By permission of Motorola Semiconductors, Inc.*)

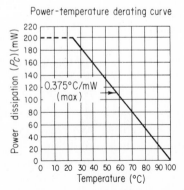

FIG. 4-7. Power-derating curve.

power-derating curve. This shows us that at the higher temperatures, the maximum collector dissipation P_C must be reduced from the normal rating at 25°C to some lower value. Again, in Chap. 11 we shall deal with this curve in detail.

QUESTIONS AND PROBLEMS

For each of the multiple-choice questions below, select the correct answer.

4-1 A certain transistor is described as having a range of β from 40 to 200. What is the range of α?
(a) 1 to 160   (b) 40 to 100   (c) 0.976 to 0.995
(d) 0.995 to 0.999   (e) 0.945 to 0.975

4-2 A transistor is off when
(a) I_C is midway between maximum and minimum value
(b) I_E is slightly greater than I_C
(c) I_B is zero and I_C is maximum
(d) I_B is zero and I_E is zero

4-3 A transistor is in the active region when
(a) $I_B = I_C$ (b) $I_E = I_C$
(c) $I_B = \beta I_C$ (d) $\beta I_B = I_C$

4-4 A transistor is saturated when
(a) the collector current is minimum
(b) $V_{CE} \cong 0$ V
(c) the emitter current is zero
(d) an increase in base current results in a corresponding increase in collector current

4-5 True or false: When operating a transistor in the saturation region, it is possible to extract a linearly amplified signal.

4-6 In the circuit below, the condition of Q_1 is
(a) saturated   (b) cut off
(c) in the active region   (d) none of these

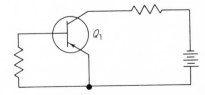

4-7 A transistor is made to increase its base current from 60 to 70 μA. It is noted that the collector current increases from 0.5 to 0.9 mA. The β_{ac} of the transistor is
(a) 0.975   (b) 0.944
(c) 40   (d) 16.7

4-8 On the curves of Fig. 4-1, if $V_{CE} = 8$ V and $I_C = 6.4$ mA, what is the value of I_B?

4-9 On the curves of Fig. 4-1, if $V_{CE} = 10$ V and I_B is 100 μA, what is the value of I_E?

4-10 On the curves of Fig. 4-1, what is the value of β_{dc} if I_C is 4 mA and $V_{CE} = 3$ V?

4-11 Using the collector characteristic curves of Fig. 4-1, the point on the curves described by $I_C = 5$ mA, $I_B = 120$ μA, $V_{CE} = 0.2$ V is in
(a) the active region
(b) the cutoff region
(c) the saturation region

4-12 The dc beta of the transistor described by the curves of Fig. 4-1 at the point where $I_B = 40$ μA and $V_{CE} = 5.5$ V is
(a) 40   (b) 50   (c) 100

4-13 Referring to Fig. 4-1, if V_{CE} is 1 V and I_B is less than 120 μA, is the transistor capable of amplifying?
(a) Yes   (b) No

4-14 Referring to Fig. 4-2a, the current through the 1-kΩ resistor is
(a) 1 mA   (b) 3 mA   (c) 6 mA

4-15 In Fig. 4-2b, the current through the 1-kΩ resistor is
(a) 6 mA   (b) 3 mA   (c) 0

4-16 In Fig. 4-2c, the current through the 1-kΩ resistor is
(a) 6 mA   (b) 3 mA   (c) 1 mA   (d) 0

4-17 In Fig. 4-4, if the transistor at room temperature has a collector current of 1.7 mA, V_{BE} is
(a) 0.6 V   (b) 0.75 V   (c) 0.35 V

CHAPTER 5 TRANSISTOR CIRCUITS

5-1 THE BASIC TRANSISTOR CIRCUIT

Up to this point the transistor has been considered an entity by itself. Except for the necessary power supply, it has not been used in any kind of practical circuit. To be put to useful work, it will necessarily be placed in some kind of typical circuit. In Chap. 3 a transistor was found to be capable of current amplification. Other attributes of a transistor circuit are equally useful, such as voltage amplification, impedance transformation, oscillation, frequency division, and so on. The results that a transistor produces are as dependent on the circuit provided for it as on the characteristics of the transistor itself.

The study of transistor circuitry finds three possible ways to connect a unit into a circuit, and these are defined by identifying the element at signal ground. These three possible circuit configurations are shown in Fig. 5-1. Each is labeled with its name, and the name is derived from the element of the transistor that is at signal ground. That is, the element that is common to both input and output is at signal ground, and this is the element from which the name of the circuit is derived. The power supplied, as well as the necessary resistors, capacitors, etc., are not shown here. The common-emitter circuit is abbreviated *CE;* the common-base *CB;* the common-collector *CC.*

In Fig. 5-2 a practical example of each circuit is shown. The common-base circuit is shown in Fig. 5-2*a*. The signal is presented to the emitter, while it is taken from the collector, with the base at signal ground. The common-emitter circuit *b* has the input presented to the base, and the output taken from the collector, while the emitter is at signal ground. The common-collector circuit *c* has the collector at signal ground, with the input presented to the base and the output taken from the emitter.

Each of these reacts quite differently to a signal, and each has specific characteristics that make it applicable to a special use. Some of the important circuit qualities are enumerated in the following discussion. Later they will be discussed in detail.

The common-base circuit is most useful for high-frequency applications. It provides voltage amplification but no current amplifi-

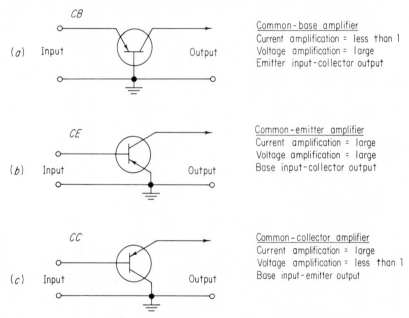

(a)   Input   Output

Common-base amplifier
Current amplification = less than 1
Voltage amplification = large
Emitter input-collector output

(b)   Input   Output

Common-emitter amplifier
Current amplification = large
Voltage amplification = large
Base input-collector output

(c)   Input   Output

Common-collector amplifier
Current amplification = large
Voltage amplification = less than 1
Base input-emitter output

FIG. 5-1.   Basic circuit configurations.

cation (or gain), and it does not invert the signal voltage. That is, a positive-going signal voltage impressed upon the emitter will result in an amplified positive-going signal voltage at the collector.

The common-emitter circuit is useful for low- or medium-frequency applications. It provides both voltage and current amplification, and it *does* invert the input signal voltage. A positive-going change on the base results in a negative-going change on the collector.

Finally, the common-collector circuit is useful at both high and low frequencies. It does not provide voltage amplification, but does give considerable current gain; it does not invert the input voltage.

5-2 BIAS

As briefly mentioned, one of the primary requirements of a transistor circuit is that it provide the proper quiescent currents and voltages.

The bias circuit is that part of the overall circuit that provides these quiescent values for the transistor.

   Bias circuits are divided into two general classes. In one of these the values of the quiescent voltages are a function of the beta of the transistor. In the other, the external circuits determine the quiescent

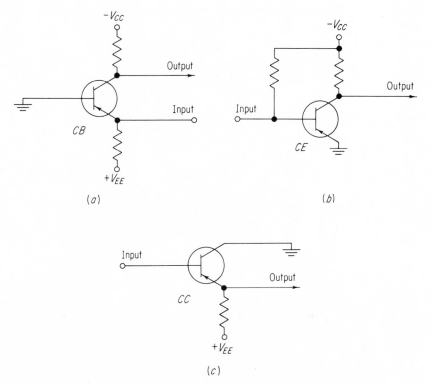

FIG. 5-2.   Practical circuit configurations.

values, and the beta of the transistor has little to do with it, provided the circuit is well designed.

CE Bias

The simplest form of biasing a transistor, often called *fixed bias*, is shown in Fig. 5-3. This is, of course, the common-emitter configuration. V_{CC} is one terminal of the collector supply voltage, RB is the resistor through which the base current flows, RL is the load resistor, and ground is the other terminal of the power supply, as shown in Fig. 5-3c.

The emitter-base junction is forward-biased by the current through *RB*. Figure 5-3*b* shows an equivalent diode circuit that acts much like part of the transistor circuit. Current flows through *RB* and through the junction, and the injection of current into the base forward-biases it, turning the transistor on. The drop across this junction is typically 0.1 V.

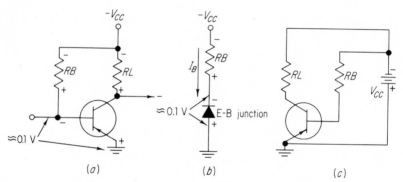

FIG. 5-3.   Biasing the common-emitter circuit.

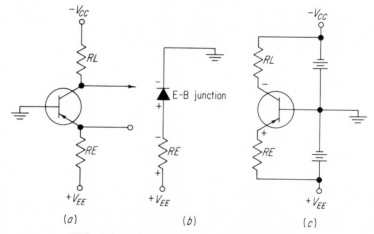

FIG. 5-4.   Biasing the common-base circuit.

With the transistor on, collector current flows, and a drop appears across *RL*. This places the collector voltage V_C, measured to ground, something more positive than $-V_{CC}$.

In this kind of circuit it is desirable to have about half of V_{CC} across the load resistor and half across the transistor. The value of *RB* sets the amount of the base current, which in turn sets I_C. Since

$I_C = \beta I_B$, the base current is very small compared with I_C. Thus RB is usually a large resistor, typically 100 kΩ, while RL may be on the order of 1 kΩ.

CB Bias

In the common-base circuit shown in Fig. 5-4, the emitter and collector resistors set the operating condition for the transistor. The base-emitter junction is forward-biased because the base is returned to a voltage more negative than that to which the emitter is returned. The amount of base current is $I_E/(\beta + 1)$. The amount of emitter current is determined by the emitter resistor RE. Again, for proper operation, about one-half V_{CC} should appear across RL.

CC Bias

The common-collector circuit shown in Fig. 5-5 is biased in a manner similar to the common-emitter circuit. Here the load resistor is in the

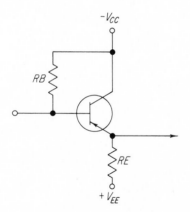

FIG. 5-5. Biasing the common-collector circuit.

emitter circuit rather than in the collector, but otherwise the two circuits are the same with respect to the dc conditions. The base current is mainly set by the value of RB. The emitter current I_E is $(\beta + 1)$ times greater than I_B. Usually, the emitter voltage V_E, referred to ground, is made to be about halfway between the two supply voltages.

Typical Bias Circuits

Figure 5-6 shows most of the commonly used bias circuits. We shall discuss each one of these only briefly here, covering them in more detail later.

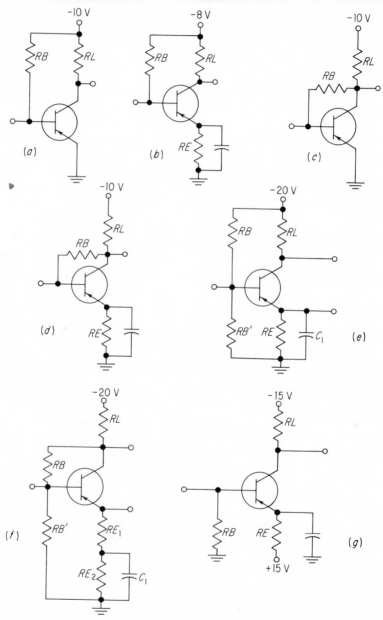

FIG. 5-6.   The seven basic biasing circuits.

Circuit a is the fixed-bias circuit, and has been briefly discussed. Circuit b is similar except for the dc negative feedback produced by RE. Circuit c is the self-bias circuit that provides negative feedback from collector to base. Circuit d is similar to c except for additional feedback via RE. Circuit e is the universal circuit that utilizes a voltage divider to set the base voltage, while circuit f is the same except for feedback from the collector. Finally, circuit g is the universal circuit using two power supplies.

All these are shown in the common-emitter configuration, although some of them can be used in other configurations.

5-3 BASIC IDEAS OF SIGNAL CONDITIONS

When a transistor is to be used in an amplifier circuit, there are many things to consider. Our present purpose, then, is to introduce these

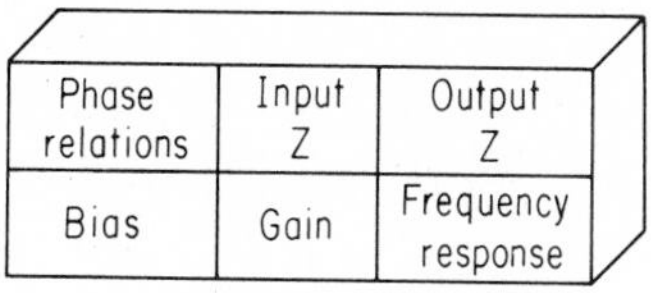

Phase relations	Input Z	Output Z
Bias	Gain	Frequency response

FIG. 5-7. Amplifier building blocks.

items, with a few preliminary statements about each. In later chapters they will be discussed in some detail. Figure 5-7 illustrates the six basic building blocks of an amplifier circuit. Each of them must be considered before an amplifier can be understood. The following points will be considered (among others) in the rest of this book.

Bias   How is the transistor biased? Is it biased for class A, class B, or class C operation? Is the amplifier biased for large or small signals, and will a change in temperature cause a change in operation?

Gain   Is the circuit to provide voltage or current amplification or both? How much will it amplify?

Frequency Response Is the amplifier to provide gain at high or low frequencies? Is it to be responsive to a single frequency or to a broad band of frequencies?

Phase Relations Will the amplifier invert the input signal?

Input Impedance Is the input impedance to be high, intermediate, or low?

Output Impedance   Is the output impedance to be high, inter-
                      mediate, or low?

When the answers to these questions can be derived for any
typical circuit, one is well on the way to a good understanding of
transistor circuitry.

5-4 AMPLIFIER-CIRCUIT CHARACTERISTICS

There are two major classes of amplifiers which concern us. The
first of these is called a *linear amplifier*. The major characteristics of

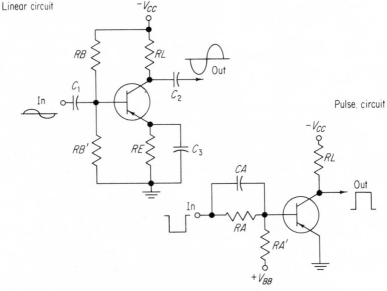

FIG. 5-8.   Differences between the linear and the pulse circuits.

the linear amplifier are: (1) the waveshape of the output signal is
an exact replica of the input; (2) the transistor is usually biased near
the center of its operating limits; (3) the transistor is never driven to
either cutoff or saturation.

The second type of amplifier is a pulse, or switching, amplifier,
and it differs considerably from the linear amplifier. Some of its
characteristics are: (1) the output is not necessarily a replica of the

input; (2) it is usually biased at either cutoff or saturation; (3) the transistor is always driven to either cutoff or saturation or both.

Figure 5-8 shows an example of each type of amplifier. Note the different circuit configurations, each being especially suited to the particular job to be done. Since we are primarily concerned with linear circuits, we shall not discuss pulse circuits.[1]

Linear-amplifier Characteristics

In Fig. 5-9 we show a typical linear-amplifier circuit. We shall now discuss several general characteristics of this circuit. First, we note that it is a common-emitter circuit. We shall expect, then, to see an inverted and amplified replica of the signal at the output, as shown. Note the indicated currents. These are listed in the table below, with the legends used in the circuit drawing.

I_{RB} The Current Through RB and RB' ————————————————

I_B Base Current ••

I_{cbo} Reverse Saturation Current — • —— • —— • —— • —— • —— •

I_C Collector Current —— - - —— - - —— - - —— - - —— - - ——

I_E Emitter Current —— - —— - —— - —— - —— - —— - ——

I_S Signal Current —— —— —— —— —— —— —— —— —— ——

I_{RB} is the resistive current drawn by RB and RB'. This current does not in any way directly influence the transistor. However, the voltage developed by I_{RB}' does set the base voltage, and so helps to determine how the transistor is biased.

I_B, the base current, flows from $-V_{CC}$ through RB to the base, where it joins with collector current to become the emitter current.

I_{cbo}, the reverse saturation current, is that part of I_C that is temperature-dependent. That is, as the temperature is increased, I_{cbo} increases, and can upset the normal bias. In the circuit shown, I_{cbo} has only a small effect within reasonable temperature variations. (See Chap. 11 for a more complete discussion of I_{cbo}.)

[1] For a complete discussion of pulse circuits, see Henry C. Veatch, "Pulse and Switching Circuit Action," McGraw-Hill Book Company, New York, 1971.

I_C and I_E are the normal collector and emitter currents. As can be seen, I_E is the sum of I_C and I_B.

I_S, at the input, is the signal current provided by an outside source, and is the current that is to be amplified. Note that I_S splits, with part going through RB', part through RB, and the remainder

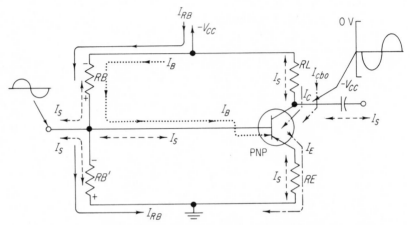

FIG.   5-9. Transistor currents, PNP.

going through the emitter-base junction. Only the part that is injected into the base can be amplified; the current flowing in the two resistors is lost as far as amplification is concerned.

Since the base is driven slightly more positive than the quiescent voltage by the signal, the base becomes more positive than before, and the transistor turns off slightly. That is, collector current decreases somewhat. With less current through RL the drop across it decreases, and the collector voltage V_C goes toward $-V_{CC}$. When the input goes negative, the collector current increases, and V_C drops toward ground. The phase inversion is clearly seen. Note that the collector never falls completely to ground or rises fully to $-V_{CC}$. Thus the input never drives the transistor full on or full off.

Figure 5-10 shows the identical circuit using an NPN transistor. Except for the polarity of V_{CC} and the direction of the direct currents, the circuit is explained exactly the same as the PNP counterpart.

Component Function

In Fig. 5-11 we show a typical linear amplifier. We shall now describe the component function and analyze some of the dc conditions in the circuit. The components shown can be described in general terms.

RL is the *load* resistor, and functions to provide a voltage drop across itself which is proportional to the current through it. Since the amplified current must flow through *RL*, the voltage drop is a replica (amplified) of the input signal.

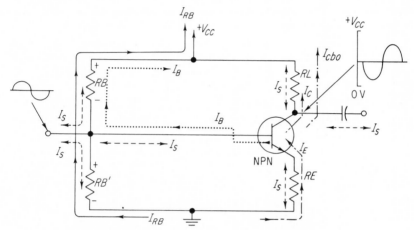

FIG. 5-10.   Transistor currents, NPN.

RB, the base resistor, along with *RB'*, assists in applying the proper dc, or quiescent, voltage (bias) to the base.

RE, the emitter resistor, helps to set the dc voltage condition. It also stabilizes the circuit so that minor changes in the transistor's characteristics have little effect on the dc condition.

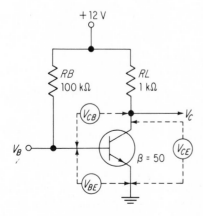

FIG. 5-11.   Fixed-bias circuit for analysis example, using an NPN germanium transistor.

Finally, C_E, the emitter-bypass capacitor, simply ensures that the voltage at the emitter cannot change rapidly; i.e., it acts as a filter to keep V_E constant, even though the base current, and therefore the emitter current, is varying.

Circuit Analysis Principles

In order to understand the many facets of transistor operation, it is of value to be able to calculate the dc (quiescent) values of voltage and current in certain simple circuits. To accomplish this, one must first be aware that the kind of circuit dictates the approach to be used. One must first locate the most logical starting point, in order to most easily determine one of the transistor currents, from which all others can be derived. In Fig. 5-6, circuits a, e, and g are the simplest to analyze; hence they will be used as examples at this point.

The fixed-bias circuit, illustrated in Fig. 5-11, is our first example. To determine all the quiescent circuit conditions, either base, collector, or emitter current must first be found. Once this is determined, the other two values may be calculated. Because this is a beta-dependent circuit, both collector and emitter currents depend upon βI_B, and hence cannot be determined first. This, of course, leaves only base current. The current flowing through RB *is* the base current, so we need only find the voltage across RB to determine the current through it. By inspection, it can be seen that, except for the base-emitter drop, the full supply voltage appears across the base resistor. (Assume $V_{BE} = 0.1$ V and $\beta = 50$.)

$$E_{RB} = |V_{CC}| - V_{BE} = 12 - 0.1 = 11.9 \text{ V}$$

The current through this resistor may now be easily found.

$$I_B = \frac{|V_{CC}| - V_{BE}}{R_B} = \frac{11.9}{100 \text{ k}\Omega} = 119 \ \mu\text{A}$$

Once base current is known, the collector current can be determined.

$$I_C = \beta I_B = 50 \times 119 \ \mu\text{A} = 5.95 \text{ mA}$$

Now, the voltage at the collector (V_C), referred to ground, is a function of the voltage drop across the load resistor RL and this is determined by the current through the resistor.

$$E_{RL} = I_C \times RL = 5.95 \text{ mA} \times 1000 = 5.95 \text{ V}$$

Finally, the voltage at the collector can be calculated.

$$V_C = V_{CC} - E_{RL} = 12 - 5.95 = 6.05 \text{ V}$$

Note that, if it is desired only to derive an approximate set of values, the base-emitter voltage can, in many cases, be ignored. Because it is so small, relative to the supply voltage, it will make little difference whether or not it is included. The following approximate relations will verify this.

$$I_B \cong \frac{|V_{CC}|}{R_B} = \frac{12}{100 \text{ k}\Omega} = 120 \text{ } \mu\text{A}$$

$$I_C = \beta I_B = 50 \times 100 \text{ } \mu\text{A} = 6 \text{ mA}$$

$$E_{RL} = I_C \times R_L = 0.006 \times 1000 = 6 \text{ V}$$

$$V_C = -V_{CC} + E_{RL} = -12 + 6 = -6 \text{ V}$$

Note that these values differ from the more precise ones by approximately 1 percent. Since the resistors used are, most probably, ±5 percent, the calculations are more accurate than necessary. It is often desirable to simplify such determinations; hence the approximations used from this point on will be found to be adequately precise. Table 5-1 illustrates both the more exact method and the approximate method.

The second example is shown in Fig. 5-12, and is seen to be a universal-bias circuit. The approach to be used here is somewhat

TABLE 5-1.   ANALYSIS PROCEDURE OUTLINE, FIXED-BIAS CIRCUIT

Exact Method	Approximate Method
1. $I_B = \dfrac{|V_{CC}| - |V_{BE}|}{R_B}$	1. $I_B \cong \dfrac{|V_{CC}|}{R_B}$ (assume V_{BE} negligible)
2. $I_C = \beta I_B$	2. $I_C = \beta I_B$
3. $E_{RL} = I_C \times R_L$	3. $E_{RL} = I_C \times R_L$
4. $-V_C = -V_{CC} + E_{RL}$ (PNP)	4. $-V_C = -V_{CC} + E_{RL}$ (PNP)
$V_C = +V_{CC} - E_{RL}$ (NPN)	$V_C = +V_{CC} - E_{RL}$ (NPN)

different than in the previous example. The value of base current is not easy to find in this circuit. However, as will be appreciated, the emitter current can be found with relative ease. To simplify matters, the value of base-emitter voltage drop will be assumed to be negligible.

Now, I_{RB} is the current flowing down from -20 V through RB

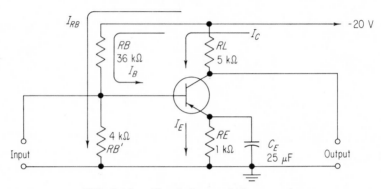

FIG. 5-12.   Circuit for analysis example.

and RB'. This current is determined by the applied voltage and the sum of RB and RB'. Note that for this calculation the base current itself is ignored by mentally disconnecting the transistor base lead.

$$I_{RB} = \frac{|V_{CC}|}{R_{\text{total}}} = \frac{20}{40 \times 10^3} = 0.5 \times 10^{-3} = 0.5 \text{ mA}$$

The voltage drop across RB' is then

$$E_{RB} = I \times R = (0.5 \times 10^{-3})\,(4 \times 10^3) = 2 \text{ V}$$

This places the junction of the two resistors at -2 V measured to ground. When we connect the base to this junction, the base will draw some current. The base current could be expected to be insignificant, perhaps 20 to 50 μA, and this would not alter the voltage by any great amount. At this point in the circuit, the voltage is determined by the resistors and the current I_{RB} through them, and is nearly independent of the base current.

If the base-emitter junction is in forward bias, the drop across the junction itself is nearly negligible, and the emitter itself is about -2 V. (The drop across the junction will be on the order of 0.1 V; so the emitter will be 0.1 V more positive than the base. The emitter, then, is actually at about -1.9 V, but for our purposes, we can say that V_E approximately -2 V.) Then the emitter current must be

$$I_E = \frac{|E_{RE}|}{RE} = \frac{2}{1 \times 10^3} = 2 \times 10^{-3} = 2 \text{ mA}$$

Since the collector current is αI_E, taking α as nearly 1, we shall make no large error if we say the collector current is nearly 2 mA.

The quiescent current through RL is such as to produce a voltage drop across it that places the collector at a more positive value than the supply. The drop across RL is

$$E_{RL} = I_C RL = (2 \times 10^{-3})(5 \times 10^3) = 10 \text{ V}$$

The collector, then, is at a potential that is 10 V more positive than -20, or referred to ground, $V_C = -10$ V.

$$-V_C = -V_{CC} + E_{RL} = -20 + 10 = -10 \text{ V}$$

The emitter resistor has a 2-V drop across it, and so the voltage across the transistor itself must be

$$V_{CE} = -V_{CC} + E_{RL} + V_E = -20 + 10 + 2 = -8 \text{ V}$$

We have defined the operating conditions for the transistor. Table 5-2 illustrates two methods of approaching this analysis problem: a more exact method and one using the various approximations for this circuit.

When a signal is applied, it will cause a deviation from these quiescent conditions. A small positive-going voltage applied to the base will partly cancel the forward bias, which results in less collector current. A small negative-going voltage applied to the base will add to the forward bias to produce a greater collector current.

The voltage drop across RL is proportional to the current through it, which is, of course, the collector current. The direction of this drop, measured to ground, is such as to be of opposite polarity to the signal voltage applied to the base. In other words, the output voltage is 180° out of phase with the input voltage. This is always true of a CE circuit. The best way to visualize this is to inspect some typical waveforms, as shown in Fig. 5-13.

The phase inversion of the CE circuit can be of importance to us in certain applications where the output of one transistor is directed

TABLE 5-2.   ANALYSIS PROCEDURE OUTLINE, UNIVERSAL-BIAS CIRCUIT

Exact Method	Approximate Method
1. $V_B = \left(\dfrac{RB'}{RB + RB'}\right)(\pm V_{CC})$	1. $V_B = \left(\dfrac{RB'}{RB + RB'}\right)(\pm V_{CC})$
2. $-V_E = (-V_B) + (V_{BE})$ (PNP) $V_E = V_B - V_{BE}$ (NPN)	2. $V_E \cong V_B$
3. $I_E = \dfrac{\lvert E_{RE}\rvert}{RE}$	3. $I_E = \dfrac{\lvert E_{RE}\rvert}{RE}$
4. $I_C = \alpha I_E$	4. $I_C \cong I_E$
5. $E_{RL} = I_C RL$	5. $E_{RL} = I_C RL$
6. $-V_C = -V_{CC} + E_{RL}$ (PNP) $+V_C = V_{CC} - E_{RL}$ (NPN)	6. $-V_c = -V_{CC} + E_{RL}$ (PNP) $V_c = V_{CC} - E_{RL}$ (NPN)
7. $\lvert V_{CE}\rvert = \lvert V_{CC}\rvert - E_{RL} - E_{RE}$	7. $\lvert V_{CE}\rvert = \lvert V_{CC}\rvert - E_{RL} - E_{RE}$
8. $I_B = I_C/\beta$	8. $I_B = I_C/\beta$

to the input of another. As one transistor is experiencing an increase in collector current, the other one will have a decreasing collector current.

When two or more linear amplifiers are used together, with the output of the first fed into the input of the second, they are said to be *cascaded.* Cascaded amplifiers work exactly the same as single circuits, except that each will affect the other somewhat. Later we shall discuss this interaction in some detail.

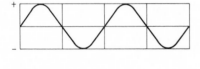

Input voltage to the base, referred to ground

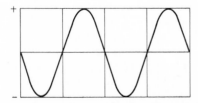

Output voltage at the collector, referred to ground

FIG. 5-13.   Common-emitter phase relations.

The final example is given in Fig. 5-14, and this is seen to be a dual-supply circuit. A still different approach must be used for this circuit to determine the quiescent values. As in the universal-circuit example, the emitter current is the logical place to start in our analysis. Note first that the base of the transistor is returned to ground

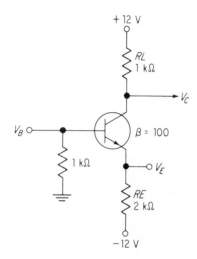

FIG. 5-14. Dual-supply circuit for analysis example, using germanium NPN transistor.

through a very low-value resistor. Because base current may be expected to be on the order of a few microamperes, the drop across this resistor can be no more than a few millivolts at most. This, relative to the total 24 V, can certainly be considered to be negligible. Hence, the base voltage can reasonably be expected to be essentially 0 V. Assuming the base-to-emitter voltage for this germanium transistor to be 0.1 V, the emitter voltage can be accurately estimated.

$$V_E = V_B \pm V_{BE} = 0.0 - 0.1 = -0.1 \text{ V}$$

The voltage across the emitter resistor allows us to find the value of emitter current.

$$E_{RE} = +V_{EE} - V_{BE} = +12 - 0.1 = 11.9 \text{ V}$$

$$I_E = \frac{E_{RE}}{RE} = \frac{11.9}{2000} = 5.95 \text{ mA}$$

Collector current is smaller than emitter current by the factor alpha.

$$I_C = \alpha I_E = \left(\frac{\beta}{\beta + 1}\right)I_E = 0.99 \times 0.00595 = 5.89 \text{ mA}$$

The drop across the collector load resistor will allow the value of V_C to be found.

$$E_{RL} = I_C \times RL = 0.00589 \times 1000 = 5.89 \text{ V}$$

$$V_C = V_{CC} - E_{RL} = 12 - 5.89 = 6.11 \text{ V}$$

As before, the analysis can be somewhat simplified by ignoring the base-to-emitter voltage drop with the following results.

$$V_E \cong 0.0 \text{ V}$$

$$I_E \cong \frac{V_{EE}}{RE} = \frac{12}{2000} = 0.006 = 6 \text{ mA}$$

Because α is very nearly 1, we can assume that $I_C = I_E$

$$E_{RL} = I_C \times RL = 0.006 \times 1000 = 6 \text{ V}$$

$$V_C = V_{CC} - E_{RL} = 12 - 6 = 6 \text{ V}$$

As before, the error can be considered to be negligible, if no better than ± 5 percent resistors are being used. Table 5-3 lists both the

TABLE 5-3. ANALYSIS PROCEDURE OUTLINE, DUAL-SUPPLY CIRCUIT

Exact Method	Approximate Method
1. $V_B = I_B \times RB \approx 0$ V (May be verified later)	1. $V_B = 0$
2. $\|V_E\| = (I_B \times RB) \pm V_{BE}$	2. $V_E \cong V_B = 0$
3. $I_E = \dfrac{\|V_{EE}\| - [(I_B RB) + V_{BE}]}{RE}$	3. $I_E \cong \dfrac{V_{EE}}{RE}$
4. $I_C = \alpha I_E$	4. $I_C \cong I_E$
5. $E_{RL} = I_C \times RL$	5. $E_{RL} = I_C \times RL$
6. $-V_C = -V_{CC} + E_{RL}$ (PNP) $V_C = V_{CC} - E_{RL}$ (NPN)	6. $-V_C = -V_{CC} + E_{RL}$ (PNP) $V_C = V_{CC} - E_{RL}$ (NPN)
7. $V_{CE} = \|V_{CC}\| - E_{RL} - E_{RE}$	7. $V_{CE} = \|V_{CC}\| - E_{RL} - E_{RE}$
8. $I_B = I_C/\beta$	8. $I_B = I_C/\beta$
9. $\|V_B\| = I_B \times RB$	9. $\|V_B\| = I_B \times RB$

approximate and more exact methods of solving for dc values in such a circuit.

5-5 COUPLING METHODS

There are three coupling methods in wide use in present-day circuitry. Although it is not our purpose to describe in detail any components other than semiconductors, a brief discussion of how these components affect the transistor amplifier is definitely in order. We shall briefly consider RC coupling, transformer coupling, and direct-coupling methods of transistor circuits.

RC Coupling

Resistance-capacitance coupling is by far the most popular choice of all coupling methods. Its chief advantages are low cost, wide fre-

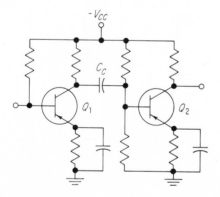

FIG. 5-15. RC coupling.

quency response, small size and weight, and general versatility. About the only disadvantages are the lack of impedance-matching qualities and the inability to select one frequency and reject all others. Where maximum concern is not directed at either of these qualities, RC coupling is used.

A typical cascaded amplifier is shown in Fig. 5-15. Q_1 is the driver stage, while Q_2 is the output stage. Q_1 is said to be coupled to Q_2 by means of an RC coupling circuit.

The coupling capacitor C_C is a limiting factor as far as the low-frequency response of the circuit is concerned. As the signal frequency becomes lower, the reactance of the capacitor becomes higher, and at some frequency will cause a voltage drop to appear across this reactance. Any voltage drop across the capacitor is a loss, since the input circuit of Q_2 will not receive this part of the total signal. Thus the reactance of the coupling capacitor and its relationship to the input resistance of Q_2 will set some lower-frequency limit for the circuit.

At the middle- and higher-frequency ranges, X_C becomes so small that the capacitor can be considered to be a short circuit to the signal frequencies. At the higher frequencies, the limiting factors are the shunt and stray capacitance of the circuit and the frequency characteristics of the transistor itself. In a later chapter we shall determine some approximations for both the low- and high-frequency limits of a typical circuit.

Transformer Coupling

When it becomes necessary to match the output impedance of a transistor to an ultimate load impedance that is either much higher

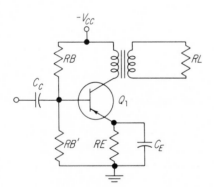

FIG. 5-16. Transformer coupling.

or much lower in value, transformer coupling is called for. This method of coupling has the advantages of good power transfer and good impedance matching. Its main disadvantages are size, weight, cost, the relatively poor overall frequency response for iron-core transformers, and often the lack of proper turns ratios available.

Figure 5-16 shows a typical transformer-coupled stage of amplification. RL is the ultimate load resistance, or impedance, and may represent the input resistance of the following stage of amplification, a loudspeaker, or any other load device requiring large amounts of current.

Usually, a step-down transformer is used to match the low-impedance load to the much higher impedance of the transistor. A numerical example will serve to show the matching qualities of the transformer.

Assume that the transistor requires a 500-Ω load for maximum power transfer. The load is to be a resistor of 8 Ω. The transformer, by virtue of the proper turns ratio, must cause the 8-Ω resistor to appear to be a 500-Ω load when viewed from the collector of the transistor.

To determine the turns ratio of the transformer, we use the following relationship:

$$\frac{R_{\text{refl}}}{RL} = \left(\frac{Np}{Ns}\right)^2$$

where R_{refl} = reflected Z
   RL = true load
   Np = number of turns, primary
   Ns = number of turns, secondary

Eliminating the exponent from the unknown quantity Np/Ns gives

$$\frac{Np}{Ns} = \sqrt{\frac{R_{\text{refl}}}{RL}} = \sqrt{\frac{500}{8}} = \frac{7.9}{1}$$

Thus the primary of the transformer must have 7.9 times as many turns as the secondary, and an 8-Ω load will appear to be a 500-Ω load to the transistor.

In many cases one frequency must be amplified while all others are rejected. This is, of course, the case of a radio-frequency amplifier, and the transformer (air core) is made to resonate by adding a capacitor to one or more windings. A later chapter will deal with this kind of circuit.

Direct Coupling

Circuits that require low-frequency response down to zero frequency (dc) must be direct-coupled. However, many problems arise when this method of coupling is used, not the least of which is a generally poorer temperature response of the overall circuit. Temperature changes will affect the circuit very slowly. So, with RC or transformer coupling a change in one stage of amplification does not cause a corresponding change in the next stage. The reactive component, of course, will not pass the dc component of the current. Figure 5-17 illustrates a direct-coupled circuit.

When circuits are direct-coupled, however, such changes are felt by the circuit and will be amplified by all succeeding stages. This causes a *drift*, or change in the operating point, of the output stages, often driving them to saturation. For this reason, direct coupling is

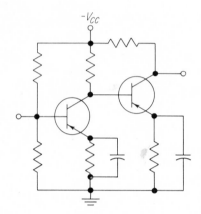

FIG. 5-17.   Direct coupling.

generally used only where necessary. Very elaborate circuits have been designed to stabilize direct-coupled circuits.

5-6 CIRCUIT LIMITS

In order to provide the user with information in the most compact form, the transistor manufacturer often supplies the characteristic curves for the product. This is quite helpful, allowing one to ascertain certain facts about the transistor itself. However, the moment the transistor is *placed in a circuit*, the curves no longer apply completely. The reason for this is simply that now the circuit imposes certain limits

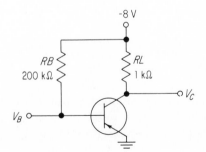

FIG. 5-18.   Circuit-limits example.

of operation, beyond which the transistor cannot go. To restore the usefulness of the curves, certain information relating to the circuit itself must be added to the curves, to specify these allowed and disallowed areas of operation. One way of combining the characteristics of the transistor and the circuit in which it is placed is to draw a dc load line upon the curves.

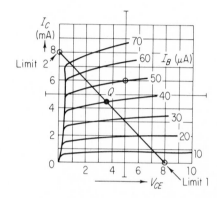

FIG. 5-19. Load line for circuit-limits example.

A dc load line is shown added to the curves in Fig. 5-19, and the circuit it represents is shown in Fig. 5-18. As has been suggested, the curves supply information relating to *all* areas of operation attainable by the transistor. For instance, the point encircled at an I_C of 6 mA and an I_B of 50 μA is a possible point of operation for this transistor. But with this circuit it is impossible to have the transistor operate at this point under dc conditions. The only values of collector current, base current, and collector-to-emitter voltage that are attainable by this transistor *in this circuit* are values that lie along this load line. Thus the load line superimposes some circuit information on the curves to indicate how the circuit and the transistor work together.

The load line is a device that allows one to see the limits of circuit operation that the load resistor and the power supply impose upon the transistor. For instance, if a transistor rated for a maximum of 30 V is used with a power supply of 8 V, the collector-to-emitter voltage can never be greater than 8 V. This is a limit that the power supply imposes upon the transistor, and even though the curves might include values up to 30 V, only values from 0 to 8 V are applicable. This maximum value of collector-to-emitter voltage is in no way related to the transistor itself, and is determined solely by the output of the power supply.

On the other hand, with some particular load resistance and supply voltage, the maximum collector current that can flow is set by simple Ohm's law, and can never exceed V_{CC}/R_{total}. This is a value set by the circuit constants. Again it is in no way related to the transistor itself.

These two circuit limits are defined as $V_{CE(max)}$ and $I_{C(max)}$. For the circuit shown these values are easily computed.

$$V_{CE(max)} = V_{CC} = 8 \text{ V}$$

$$I_{C(max)} = \frac{V_{CC}}{RL} = \frac{8}{1000} = 0.008 \text{ A, or 8 mA}$$

The DC Load Line

To construct the dc load line of Fig. 5-19, simply locate the two limits upon the proper ordinate of the graph at V_{CE} equal to 8 V (limit 1) and I_C equal to 8 mA (limit 2). A straight line drawn between these two points is the dc load line. The transistor can be operated only at values of voltage and current that fall along this line. Any other combinations of dc voltage and current are impossible for the transistor to attain in the circuit shown.

One of the most important uses for the load line is as an aid in determining the quiescent operating point, or Q point. The Q point is specified by the steady-state condition (no signal applied) of collector-to-emitter voltage, collector current, and base current. This is very important to know, as will be seen later, and is the first thing that must be determined when making a complete circuit analysis.

There is one point along the load line that will specify the three values, V_{CE}, I_B, and I_C, that apply to the circuit under dc conditions. Let us determine this one point, the Q point, for this circuit.

The first step in analyzing such a circuit is to locate some value of current that can be readily and easily determined. In this circuit, I_B is determined by the values of V_{CC} and RB. I_C and I_E, on the other hand, depend upon both beta and I_B; so they cannot be determined first. The equivalent resistance of the forward-biased base junction is very low, and compared with 200 kΩ, can be considered negligible. Thus I_B is very nearly equal to V_{CC} divided by RB.

$$I_B = \frac{V_{CC}}{RB} = \frac{8}{200 \text{ k}\Omega} = 40 \text{ }\mu\text{A}$$

On the curves of Fig. 5-17 the intersection of the 40-μA base-current line and the load line is the Q point, as shown. Quiescently (with no signal), the base current must be 40 μA, with I_C equal to about 4.4 mA, while V_{CE} is about 3.7 V. Thus, if we have a set of curves that exactly describes our transistor, we can easily determine the dc operating point of the circuit.

In actual practice, the curves supplied by the manufacturer are average curves and give only typical values. Since transistor characteristics vary so widely, the only way to get accurate curves for a particular transistor is to use an oscilloscope curve tracer and obtain a photograph of the actual curves for this transistor. Then the values given by the curves will apply only to the unit in question.

Base-Emitter Resistance

The statement was made previously that the forward-biased emitter junction represented very little resistance compared with RB. We shall now verify that this was a reasonable assumption for germanium junction transistors. The actual ohmic, or bulk, resistance of the emitter material we call R_{EE}. This is the opposition afforded to current as emitter current flows from the base region across the emitter material itself. For a junction transistor, the value of R_{EE} is best determined from the manufacturer's input curves, such as were shown in Figs. 4-3 and 4-4. If the curves are not available, the value of R_{EE} can be estimated rather closely by using the following approximations. The equations below are reasonably valid for all junction transistors (except the power types) that have collector currents of less than about 20 mA.

$$R_{EE} \cong \frac{100}{I_E(\text{mA})} \qquad \text{for germanium units}$$

$$R_{EE} \cong \frac{700}{I_E(\text{mA})} \qquad \text{for silicon units}$$

It should be kept in mind that R_{EE} is the emitter resistance to direct currents only. As an example of finding the emitter resistance to the dc bias currents, assume that a germanium transistor is being used in a circuit that produces a collector current of 1 mA.

$$R_{EE} \cong \frac{100}{I_E} = \frac{100}{1} = 100 \ \Omega$$

It will now be shown that R_{EE}, when viewed from the base, is not the same as when viewed from the emitter. As explained below, R_{EE}, when viewed from the base, is $(\beta + 1)$ times greater than the ohmic value. If beta is 100, the example above will yield a value of about 10,000 Ω. Because this is so different, we give it a special name, and call it R_{BE}.

The fact that R_{BE} is $(\beta + 1)$ greater than R_{EE} can be explained as follows: If, in the emitter region, the current density is, for example, 1 mA and base current is 10 μA, then the relative resistances must be different. The same voltage must be across the junction whether it is viewed from the base or from the emitter. Thus we have a situation similar to the case of two parallel resistors through which different values of current are flowing. Suppose the voltage across two parallel resistors is 100 mV. If one of them is carrying 1 mA, its resistance must be E/I, or $0.10/0.001 = 100$ Ω. If the current through the other resistor is 10 μA, its resistance is also E/I, but now we must

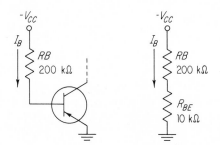

FIG. 5-20. Base-emitter resistance.

use the new values. $E/I = 0.10/0.00001 = 10,000$ Ω. This is almost exactly the situation that exists at the emitter-base junction of a transistor. Since the base current itself must flow across the emitter, we have not $\beta \times R_{EE}$, but $(\beta + 1) \times R_{EE}$ as the equivalent resistance when viewed from the base. In the example above, then, the actual value of R_{BE} is 10,100 Ω.

Thus the base-emitter region of the transistor will have two different values of resistance: R_{EE} is the value viewed from the emitter (that is, in terms of emitter current), while R_{BE} is the value viewed from the base (in terms of base current).

When determining the value of base current, as in the original problem, the equivalent base resistance in series with the base resistor RB is R_{BE}. Figure 5-20 shows an equivalent circuit with RB equal to 200 kΩ and R_{BE} equal to about 10,000 Ω. This would be calculated as shown below.

Assume $I_E = 1$ mA.

$$R_{EE} = \frac{100}{I_E} = \frac{100}{1} = 100 \ \Omega$$

$$R_{BE} = (\beta + 1) R_{EE} = 101 \times 100 = 10.1 \ \text{k}\Omega$$

It can now be seen that 10,100 Ω is relatively small compared with 200 kΩ, and that it can be safely ignored as far as its influence on base current is concerned, if precise accuracy is not required.

DC Load Line with Emitter Resistance

With somewhat more complex circuits, the load line is nearly as simple to construct as in the previous load-line example (Fig. 5-19) if the proper information is used. Consider Fig. 5-21, where we have

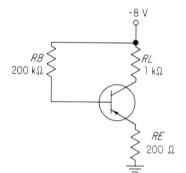

FIG. 5-21. Circuit for load-line variation.

added a resistor RE to the basic circuit. We shall analyze this circuit for its overall dc condition. The load line is constructed in much the same way as before but we must be careful to include RE where applicable. Limit 1 is still 8 V. However, limit 2 is now a different value, since there is more resistance in the collector and emitter circuits.

$$I_{C(\text{max})} = \frac{V_{CC}}{RL + RE} = \frac{8 \ \text{V}}{1.2 \ \text{k}\Omega} \cong 6.7 \ \text{mA}$$

The upper end of the load line must now be limited to 6.7 mA, rather than 8 mA as before, as indicated in Fig. 5-22.

The first step in determining the Q point of the circuit is to find the dc base current. This is not quite as simple as before, since there is the emitter resistor to consider. Looking from the base itself toward

ground, there are several values to consider. One of these is the
emitter resistor RE, a 200-Ω resistor. Looked at from the standpoint
of the base current, this resistance appears to be $(\beta + 1)$ larger than
it actually is. Also, R_{BE}, the equivalent junction resistance, is $(\beta + 1)$
larger than its actual value. At this point we can only approximate the

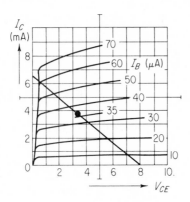

FIG. 5-22.   Load line with RE.

emitter current; so a reasonable value to assume is one-half the
maximum value, or $6.7/2 = 3.35$ mA.

Now R_{EE} can be approximated.

$$R_{EE} = 100/3.35 = 30 \ \Omega$$

This appears greater by the factor of $(\beta + 1)$. (Assume $\beta = 100$.)

$$R_{BE} = 101 \times 30 \cong 3000 \ \Omega$$

RE appears 101 times greater than 200 Ω, and the sum of these two
values is the total resistance seen by the base looking toward ground,
which we call R_{ib}.

$$R_{ib} = (\beta + 1) \ R_{EE} + (\beta + 1) \ RE = 3000 + 20.3 \ k\Omega \cong 23.3 \ k\Omega$$

An equivalent circuit can be drawn for the path of base current,
as shown in Fig. 5-23. This is a simple series circuit, and is solved for
the base current as shown.

$$I_B = \frac{V_{CC}}{R_{\text{total}}} = \frac{8 \ V}{223 \ k\Omega} \cong 36 \ \mu A$$

Since beta is 100 for this transistor, collector current must be 100 times 36 μA.

$$I_C = \beta \times I_B = 100 \times 36 \ \mu A = 3.6 \ mA$$

Because $I_C \cong I_E$, we can now determine the drop across the load resistor and the emitter resistor.

$$E_{RE} = I_E \times RE = 3.6 \ mA \times 200 = 0.72 \ V$$

$$E_{RL} = I_C \times RL \cong 3.6 \ mA \times 1000 = 3.6 \ V$$

V_C, referred to ground, is simply $-V_{CC} + E_{RL} = -8 + 3.6 = -4.4$ V, and V_E is -0.72 V.

It should be noted that, because of several simplifying assumptions, the above calculations are not exactly correct. By assuming a value for I_E at the start, we did not arrive at an exact figure. But the value of R_{BE} that was derived from this assumption is very much smaller than the value of the equivalent emitter resistor viewed from the base. Also, the assumption that I_C is nearly the same as I_E intro-

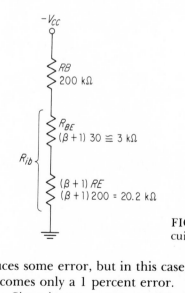

FIG. 5-23. Base equivalent circuit.

duces some error, but in this case, if beta is on the order of 100, this becomes only a 1 percent error.

Since in many cases we are using resistors with a tolerance of ±5 percent to bias the transistor, this error is small indeed. Even when ±1 percent resistors are used, this error is of the same order as the error introduced by the resistors themselves.

Once the dc conditions of a circuit are found, it becomes necessary to know something about the conditions as they vary because of
a signal. Using the circuit of Fig. 5-21, suppose a 20-μA peak-to-peak
signal is impressed upon the base.

During the half-cycle that the input aids the base current (base
going more negative), more base current flows. When the signal
opposes the base current (base going more positive), less base current
flows. This action is shown on the curves of Fig. 5-24, where point 1

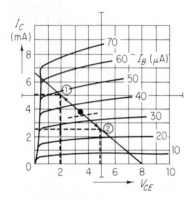

FIG. 5-24. Collector excursion
along the load line.

is maximum base current and point 2 is minimum. The collector-to-
emitter voltage swing is evident, as is the collector-current swing.
Thus, from the curves, we can visualize the action of the transistor
as a signal is impressed.

QUESTIONS AND PROBLEMS

5-1 Using the circuit of Fig. 5-18 and the curves of Fig. 5-19, determine the dc voltage at the collector if *RB* is changed to
100 kΩ.

5-2 Using the curves of Fig. 5-22 and the circuit of Fig. 5-21, draw
the load line if V_{CC} is 6 V. Locate the Q point on the load line.

5-3 Using the circuit shown in Fig. 5-21, compute the value of *RE*
as it looks from the base if $\beta = 55$.

5-4 Briefly discuss the reason why each of the circuits of Fig. 5-6 allows the emitter-base junction to be forward-biased.

5-5 Briefly discuss the direct currents shown in Fig. 5-9. Relate the magnitude of each in relative terms to each of the others.

5-6 In the circuit of Fig. 5-12, change the value of RL to 7.5 kΩ. Calculate V_B, V_E, V_C, and I_B. Beta is 75 for this example.

5-7 Discuss the reason for the voltage inversion of the signal in the common-emitter circuit.

5-8 Discuss the pros and cons of RC coupling; of transformer coupling; of direct coupling.

5-9 Refer to Fig. 5-9. The signal current is shown flowing across the base-emitter junction. Briefly discuss the reason why this is not the total signal current injected from an outside source.

5-10 Refer to Figs. 5-9 and 5-10. Discuss briefly the differences between the following currents on each drawing: I_C; I_B; I_E; I_S; I_{cbo}.

5-11 Refer to the circuit of Fig. 5-12. The resistor R_B is to be changed to 44 kΩ. Calculate the approximate value of V_B at junction of RB and RB'; V_E; E_{RL}; I_E.

5-12 Refer to Question 5-11. Which of the answers given below is correct? The operating point just determined tells us that the transistor is biased at
(a) cutoff   (b) saturation   (c) the active region

5-13 Again referring to Question 5-11, what is the value of R_{EE} if the transistor is germanium?

5-14 Refer to Figs. 5-21 and 5-22. Change RE from 200 to 400 Ω. Determine R_{EE}; R_{ib}; E_{RL}. Draw the equivalent circuit of the base-current path. Assign values to all resistances. (Transistor is germanium.)

5-15 In the circuit of Fig. 5-12, change RB to 24 kΩ. Determine V_B; V_E; E_{RL}; I_E.

5-16 Refer to Figs. 5-18 and 5-19. In the circuit, the load resistor is to be changed from 1 to 2 kΩ. Construct the new load line. Determine the Q point for the new load line.

5-17 Refer to Fig. 5-24. The base-current change is reduced from 20 μA peak to peak to 10 μA peak to peak. What is the new collector-current excursion? What is the new collector-to-emitter voltage excursion?

5-18 Refer to Fig. 5-24. What is the maximum peak-to-peak variation of base current necessary to drive this transistor circuit to the edge of saturation and to cutoff?

5-19 Using the circuit and curves shown in Figs. 5-18 and 5-19, determine the dc voltage at the collector if the base resistor RB is changed to 150 kΩ.

5-20 Using the circuit and curves shown in Figs. 5-18 and 5-19, determine the dc voltage at the collector if the base resistor is changed to 400 kΩ.

5-21 Again using Figs. 5-18 and 5-19, determine the value of RB if V_{CE} is to be 5.8 V.

5-22 Still referring to Figs. 5-18 and 5-19, RB and RL are to remain as shown, but $-V_{CC}$ is to be changed to -10 V. Determine the following values for this new condition: I_B; I_C; V_C.

5-23 A circuit of similar configuration to Fig. 5-18 is to be designed using a NPN transistor. The collector supply voltage will be 12 V. The transistor characteristics are (1) $\beta = 100$; (2) $V_{BE} = 0.75$ V. The load resistor is to be 1500 Ω.
(a) Draw the circuit, showing values known at this time.
(b) Determine the value of RB so as to make $V_{CE} = 6$ V.
(c) Determine a value of I_C that will allow V_C to be $\frac{1}{2}V_{CC}$.
(d) Determine the value of I_B flowing in step (c) above.

5-24 Refer to Fig. 5-18. The transistor has a β of 35. Determine the value of RB to make V_{CE} equal to $\frac{1}{2}V_{CC}$.

5-25 Refer to Fig. 5-18. The transistor has a β of 65. Determine the value of RB that will make V_{CE} equal to $\frac{1}{2}V_{CC}$.

5-26 Refer to Fig. 5-18. Change the values listed to the following: $RB = 100$ kΩ; $RL = 1$ kΩ; $\beta = 50$; $-V_{CC} = -8$ V. Find the value of R_{EE}.

5-27 Using the circuit values in Question 5-26, find the value of R_{ib}.

5-28 Using the circuit values in Question 5-26, determine the drop across RL.

CHAPTER 6 THE COMMON-EMITTER CIRCUIT

We have already briefly discussed the three basic circuit configurations. In the next three chapters we shall investigate each of these separately, to derive some fundamental ideas of circuit action. Our immediate concern will be to determine the dc conditions, the large-signal response, and the small-signal response of the common-emitter circuit.

One of our ultimate goals is to find a method of extracting circuit information equivalent to that obtained from the characteristic curves, but without the use of the curves. To establish a good foundation upon which to build, we shall describe many of the general attributes of the basic circuits. Then, using these concepts, we shall find a simple, fast, and accurate way to describe our circuits and to analyze for the measurable circuit values.

We shall use the *CE* circuit to explain most of the fundamental properties of transistor circuits, because most of the circuits one encounters are of this configuration. Following this chapter will be two chapters on the other configurations. Each will be analyzed in a similar manner so far as possible, considering the different characteristics.

6-1 CHARACTERISTICS OF THE COMMON-EMITTER CIRCUIT

The common-emitter circuit is the most widely used configuration because its basic characteristics fit the requirements of most circuits. A few of these characteristics are enumerated below, to provide a generalized picture of the *CE* circuit.

A typical *CE* circuit is shown in Fig. 6-1, and as indicated by the waveforms, the output voltage e_o is 180° out of phase with the input. This is a distinctive feature of this configuration. Also evident is voltage and current gain, and this, too, is a feature not shared with either of the other two circuits. The input resistance consists primarily of

R_{BE}, which is usually quite low. The equivalent input resistance is actually shunted by RB, but RB is so much larger than R_{BE} in the usual circuit that it seldom needs to be considered. Generally speaking, the high-frequency response of the CE circuit is poor. It can be made to be reasonably good, but only at the expense of voltage gain.

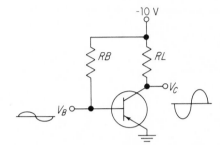

FIG. 6-1. Typical common-emitter circuit.

These are the major characteristics of the CE circuit described in very general terms. When it becomes necessary to become more specific, an analytical method must be employed to describe the circuit response in more exact mathematical terms. Thus we must have these data about a circuit: where it is operating with no signal applied, the gain, the frequency response, the large-signal response, the phase relations, and the input resistance. The basic ideas that follow will provide a foundation upon which to build the analytical method that will allow us to analyze properly almost any circuit.

6-2 METHODS OF BIASING

One of the important things to know about any circuit is the method of biasing. Several basic methods are in common use, most of which can be related to one of the seven basic configurations shown in Fig. 6-2a to g. At this point we wish merely to get acquainted with these seven basic circuits, which are divided into two groups.

The first group (a to d) is called the beta-dependent circuits, and the second (e to g), the beta-independent circuits. The beta-dependent circuits are so called because the quiescent operating point is entirely a function of the transistor's beta. In other words, two similar transistors placed in the same circuit may, and probably will, operate at entirely different points on their characteristics. One of them may be nearly at cutoff, while the other may be near saturation because of the wide variance of beta from transistor to transistor.

On the other hand, a well-designed beta-independent circuit

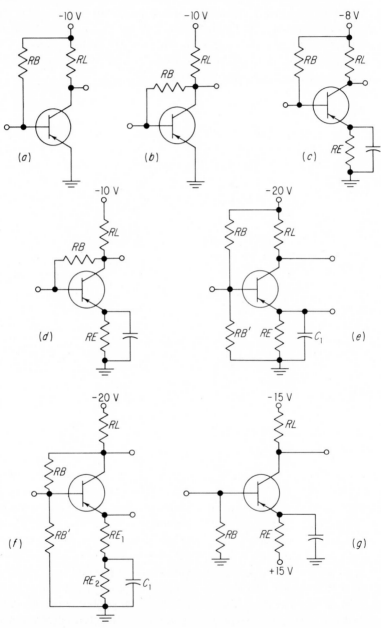

FIG. 6-2.   Beta-dependent circuits: (*a*) fixed bias; (*b*) self-bias; (*c*) fixed bias with *RE;* (*d*) self-bias with *RE.* Beta-independent circuits: (*e*) universal bias; (*f*) universal circuit with collector and emitter feedback; (*g*) dual-supply universal circuit.

operates at a fixed point regardless of the transistor's beta. Thus it is readily seen that beta-dependent circuits are seldom used, because of the difficulty of predicting results. However, enough of them exist to make it necessary to know how they work. (The voltages are shown in Fig. 6-2 to indicate typical values only.)

Beta-Dependent Biasing

In Fig. 6-2, circuit a is the simplest form of biasing a CE circuit. The base, being returned to V_{CC}, is allowed to draw base current in an amount determined by V_{CC} and RB. Thus the base-emitter junction is forward-biased. The collector current is beta times the base current. The base current is V_{CC}/RB. The circuit is very sensitive to changes in temperature.

Circuit b introduces both dc and ac feedback to the first circuit. The junction is forward-biased by the same mechanism as before. The relationship between I_B and I_C is still the same, but is not as simple to determine as before. It exhibits better temperature characteristics than circuit a, but has less voltage gain.

Circuit c has an added resistor in the emitter. Again, the relation between I_C and I_B is the same, but it is more difficult to determine than in circuit a. This circuit has better temperature response than either of the other circuits shown, without sacrificing gain if RE is bypassed.

Circuit d is a combination of circuits b and c, and exhibits good temperature stability with reduced signal gain. RE is often bypassed for signal frequencies.

Beta-Independent Biasing

Circuit e of this group is often called the *universal* circuit, because it can be used as a CE, CC, or CB amplifier. From a dc standpoint it does not make any difference what the configuration is. The base voltage V_B is set by the voltage divider in the base circuit, and this makes the circuit relatively independent of the transistor's beta. The voltage at the base, and therefore the voltage at the emitter, depends only upon the external circuit values. By proper choice of component values, the base voltage, referred to ground, is made just negative enough to forward-bias the junction. This is somewhat critical, because a voltage here that is too negative will drive the transistor into saturation. However, done properly, the transistor is easily biased at any point on the dc load line regardless of the characteristics of the transistor. The temperature stability of this circuit is excellent, and if *RE* is bypassed, the voltage gain may be made very high.

Circuit *f* is different from circuit *e* only in that the base resistor is returned to the collector rather than to V_{CC}. This produces both dc and ac feedback, resulting in better temperature stability but reduced gain. (The fact that there are two emitter resistors does not change the dc response of the circuit. This is done to improve the signal conditions.)

Circuit *g* is very closely related to circuit *e*. This circuit has all the attributes of circuit *e*, but uses one less resistor. If *RE* is bypassed, the gain can be made very high. Temperature stability is excellent.

6-3 THE Q POINT

In order to analyze properly any transistor circuit, the first thing that must be known is the dc condition of the circuit. We have been calling this the quiescent, Q, point, and this is specified by the dc voltage and current at each leg of the transistor. A meaningful signal analysis cannot be made until the Q point is specified. To prove this statement, Fig. 6-3 shows a typical *CE* circuit. One could spend quite some time rendering an ac analysis of this circuit, but this time would be wasted.

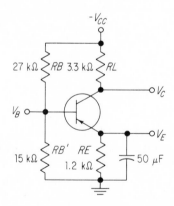

FIG. 6-3. Universal-circuit example.

This transistor is biased *in saturation,* and the usual expressions relating to gain, etc., have no meaning. The transistor is not capable of amplifying, but until a dc analysis is made, we cannot know this. Therefore the quiescent condition of the circuit must first be specified before completing any analysis.

The quiescent point of any circuit can be determined in many ways. An example will serve to show how this might be accomplished. A typical simple circuit is shown in Fig. 6-4, and we shall work our way through a simplified dc analysis of this circuit step by step. In Chap. 5 we determined the Q point of a very similar circuit with the idea of plotting this point on the load line. Now we want to determine V_C, V_B, and I_C without using the curves.

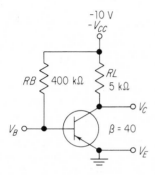

FIG. 6-4.   Fixed-bias example.

We know that in any circuit the collector current is beta times the base current. Thus the most logical starting place is the base current. In this circuit, I_B is determined by RB and V_{CC}.

$$I_B = \frac{|V_{CC}|}{RB} = \frac{10}{400 \text{ k}\Omega} = 25 \ \mu\text{A}$$

I_C, then, is beta times 25 μA. (Beta is 40.)

$$I_C = \beta I_B = 40 \times 25 \ \mu\text{A} = 1 \text{ mA}$$

Knowing the collector current now allows us to determine the drop across the load resistor, E_{RL}. This in turn will allow the collector voltage V_C, referred to ground, to be determined.

$$E_{RL} = I_C \times RL = 1 \text{ mA} \times 5 \text{ k}\Omega = 5 \text{ V}$$

$$-V_C = -V_{CC} + E_{RL} = -10 + 5 = -5 \text{ V}$$

If the emitter current is of interest, we can say that since α is nearly 1, the emitter current and the collector current are nearly equal. Usually, this is a valid assumption. Only if beta is approximately 20 or less would this have to be considered. If beta is 100 or so, the error introduced by assuming that I_C equals I_E is on the order of 1 percent. If 5 percent resistors are used, this small error can be easily tolerated. If it must be considered, it is easy enough to simply say that the emitter current is $(\beta + 1)$ greater than I_B. Then an accurate value for I_E can be used for the necessary calculations. With these few calculations, then, we have determined the dc operating conditions of this circuit.

6-4 GAMMA

Usually, the characteristic curves are consulted when it is necessary to determine the Q point of a circuit. A load line is drawn, and the proper currents and voltages are superimposed upon the curves, and these will delineate the Q point. One reason why this method has achieved such popularity is that it allows one to visualize the circuit conditions.

It is, however, possible to visualize the operating point of a circuit without using the curves themselves. The concept of the circuit gamma (γ) is very useful to describe the Q point of any transistor circuit.

The dc load line, constructed on the curves of a transistor, delineates the possible points of operation attainable by the transistor in a given circuit. The dc operating point is one point on the load line that describes the V_{CE}, I_C, and I_B of the circuit when no signal is applied. The gamma (γ) of the circuit gives us the same information without in any way using the characteristic curves.

Any operating point plotted on the load line is specified by a number of combinations of voltage and current. Any two of these values will locate one single point on the curves. Using the example of Fig. 6-4, the operating point can be located by a base current of

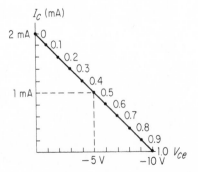

FIG. 6-5. Illustrating gamma.

25 μA and a collector current of 1 mA. Or a collector current of 1 mA and V_{CE} of 5 V will locate the same point.

The two values used to determine gamma are V_{CE}, the collector-to-emitter voltage, and V_{CC}, the collector supply voltage. In this case, V_{CE} is 5 V, while V_{CC} is 10 V. The load line is shown plotted in Fig. 6-5. The Q point is shown, and is seen to be at the center of the dc load line. The relationship of V_{CE} to V_{CC} determines the value of gamma. (The base-current curves are not shown since they do not directly influence the present discussion.)

In the example, gamma is determined as follows, using the absolute values of voltage:

$$\gamma = \frac{V_{CE}}{V_{CC}} = \frac{5}{10} = 0.5$$

This indicates that V_{CE} is one-half of V_{CC}, and *this is always the center of the dc load line.*

These particular values were chosen to define gamma because V_{CC} is always known, and V_{CE} can be easily determined. (We shall later find a very simple way to determine the circuit gamma, using different constants.) Thus, knowing V_{CE} and V_{CC} will allow us to visualize where on the load line any circuit is biased. If the circuit gamma is 1, or nearly 1, the transistor is biased at, or nearly at, cutoff. On the other hand, if gamma is about 0.1, the transistor is nearly in saturation. By simply remembering where on the load line these numbers are located, we can dispense with the curves for this determination.

A few examples will serve to illustrate the usefulness of the concept of gamma. We shall use the load line of Fig. 6-6 for illustration purposes only.

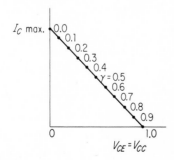

FIG. 6-6.   Gamma example.

A transistor is operating with V_{CE} equal to 20 V. V_{CC} is 25 V. We wish to find the bias condition.

$$\gamma = \frac{V_{CE}}{V_{CC}} = \frac{20}{25} = 0.8$$

The region of operation for this transistor is a point on its load line $\frac{2}{10}$ of the way up from cutoff.

Another transistor is operating with a V_{CE} of 2 V, while V_{CC} is 20 V. We wish to find where, on its load line, it is biased.

$$\gamma = \frac{V_{CE}}{V_{CC}} = \frac{2}{20} = 0.1$$

In this case, the transistor is biased very nearly in saturation, as indicated on the load line where gamma is 0.1.

With any combination of V_{CE} and V_{CC}, the quiescent point of operation along the load line can be easily and quickly determined without using the transistor curves.

6-5 TRANSISTOR DC EQUIVALENT CIRCUIT

An important concept that deviates somewhat from the more conventional approach stems from the use of an extremely simplified dc equivalent circuit of the transistor. This is not a new idea itself, but will be in the use we shall eventually make of it. Figure 6-7a is an

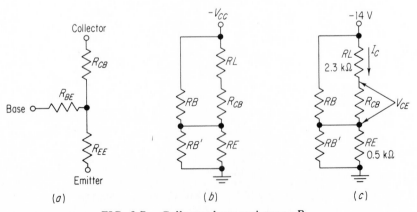

FIG. 6-7. Collector-base resistance R_{CB}.

approximate dc equivalent circuit of a transistor, showing the resistances of each junction. R_{EE} is the dc ohmic resistance of the emitter region, and for a germanium unit is defined as $0.1/I_E$, or $100/I_E$ (mA). This is often about 100 Ω or so in actual value. R_{BE} is the effective resistance, base to emitter, viewed from the base, and is defined as $(\beta + 1) \times R_{EE}$. Both of these values can be ignored in the present case, either because they are so small or because the current through the junction is very small.

On the other hand, the reverse-biased collector-base junction is a relatively large value, and cannot be ignored, because of the large current flowing through it; we call this resistance R_{CB}. An example of R_{CB} and its relation to the external circuit is shown in Fig. 6-7b. By ignoring R_{BE} and R_{EE}, we can simplify to a large degree the dc conditions in the overall circuit. That this is a valid simplification is proved when we realize that, so long as the transistor is in the active region, the voltage difference between the base and the emitter seldom exceeds 0.1 V for germanium transistors. By ignoring this difference, the circuit of Fig. 6-7b becomes a perfectly valid equivalent circuit to work with. The error introduced by considering the base and emitter to be at the same voltage is far smaller than that introduced by the tolerance of the resistors used in the circuit.

The value of R_{CB} is properly determined, as with any resistance, by the voltage drop across it and the current flow through it. Since we are ignoring the drop across the base-to-emitter junction, we can use the voltage across the entire transistor, V_{CE}. This is the same as the value derived from the curves of the transistor. Thus I_C and V_{CE} will determine the value of R_{CB}.

As an example of finding the value of R_{CB}, the circuit of Fig. 6-7c can be used.

Collector current in this circuit is 2.5 mA, and the voltage drop from the collector to the emitter is 7 V.

$$R_{CB} = \frac{V_{CE}}{I_C} = \frac{7}{2.5 \text{ mA}} = 2.8 \text{ k}\Omega$$

where V_{CE} is equal to the supply voltage less any drops across the external resistors (RL and RE).

At the quiescent operating point, the dc resistance of the reverse-biased collector junction is 2800 Ω. The importance of replacing the transistor by R_{CB} in a simplified circuit is that, since all circuit values are known *except* R_{CB}, if we can find a simple way of finding a value of R_{CB}, we can then use Ohm's law to determine all quiescent values for the circuit. We shall later find an exceptionally simple way to analyze typical circuits for their quiescent and large-signal values by using this concept of R_{CB}.

To some the foregoing simplification may seem extreme. At first glance it would seem that the transistor is more complex than this. Indeed it is! However, remember that our present purpose is to find a way to yield a simplified approach to analyzing a circuit for its measurable quantities. The concept of R_{CB} will later be found to be a useful tool that will help in understanding many of the qualities of typical circuits, some of which are quite complex. Thus we feel that, in this special case, the oversimplification is justified.

6-6 LIMITS OF THE CIRCUIT

Another very important concept that is of value in analyzing certain circuits is the circuit *limits*. By the use of this concept we can apply some of the results of graphical analysis to a method that does not use the characteristic curves themselves. As mentioned before, in connection with graphical analysis, every practical circuit has certain limits beyond which it cannot go. Normal operation, then, is between these extreme limits.

As before, there are two limits of interest, one of which is the condition with the transistor in saturation. The other limit occurs when

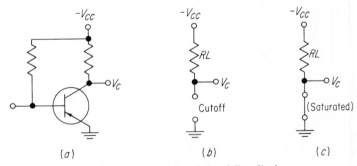

FIG. 6-8.   Simple circuit load-line limits.

the transistor is cut off. In the simple beta-dependent circuit, these limits are set only by V_{CC} and ground. In the case of the more complex circuits, other things must be considered.

We are mainly concerned here with the limits of both collector and emitter voltage. Figure 6-8 shows a simple circuit along with its two collector limit conditions, while Fig. 6-9 shows the same conditions for a more complex circuit.

The simple circuit is easy to visualize for both the cutoff and saturated conditions. If the transistor is cut off, it appears to be an open circuit, and with no current through RL there is no voltage drop

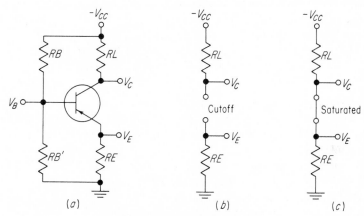

FIG. 6-9.   Universal-circuit limits.

across it. V_C, then, must equal $-V_{CC}$. If the transistor is in saturation, it appears to be a short circuit; so the full supply voltage is dropped across RL, and V_C is at ground. (We neglect the 100 mV or so that actually appears across the transistor when it is in saturation.) Thus the two limits for the collector voltage are $-V_{CC}$ and ground. V_C can go no more negative than $-V_{CC}$ and no more positive than ground. V_E is, of course, directly connected to ground, so cannot be at any other potential.

In Fig. 6-9, however, we have a different situation. The emitter is floating above ground at some value more negative than ground. Since it is not bypassed, it is free to move to any potential *within its limits*. If the transistor is in cutoff, it again appears as an open circuit. Since no current flows through R_E, there is no drop across it, and V_E is at ground potential. Also, V_C must be at $-V_{CC}$ for the same reason. The other limits, with the transistor in saturation, must be some value between $-V_{CC}$ and ground.

Now, a simple voltage divider exists that consists of RL and RE. If the transistor appears to be nearly a dead short, V_C and V_E are at nearly the same potential. Thus *the voltages at V_E and V_C are set by $-V_{CC}$ and the two resistors.*

As a numerical example, suppose $-V_{CC}$ is -20 V, and $RL = 2$ kΩ and $RE = 1$ kΩ. One of the maximum limits of V_E is ground, of course. The other limit is with the transistor saturated, and then

$$-V_{E(max)} = \frac{RE}{RL + RE}(-V_{CC}) = \frac{1 \text{ k}\Omega}{2 \text{ k}\Omega + 1 \text{ k}\Omega}(-20) \cong -6.7 \text{ V}$$

The emitter can range between ground and -6.7 V, but it can go no farther in either direction. *As long as V_E is between these limits, the transistor must be in the active region.* Thus, once the emitter limits are known, V_E is a good indication of what region the transistor is biased in.

V_C has two specific limits. Once the V_E limit is determined, we know the limits of V_C also. V_C can only range between $-V_{CC}$ and $V_{E(max)}$, which in the example is -6.7 V. As long as V_C is between these limits, the transistor must be in the active region.

How we apply these ideas of circuit limits is best shown by an example. If we analyze the circuit of Fig. 6-10, we shall be able to see the value of this concept of limits.

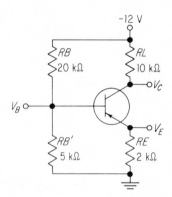

FIG. 6-10.   Circuit example.

$V_{E(min)}$ is, of course, ground. $V_{E(max)}$ is determined as before.

$$-V_{E(max)} = \frac{RE}{RL + RE}(-V_{CC}) = \frac{2 \text{ k}\Omega}{12 \text{ k}\Omega}(-12) = -2 \text{ V}$$

V_E can only be as negative as -2 V if the transistor is in saturation.

We must now determine what the base-circuit voltage divider is trying to put on the base of the transistor.

$$-V_B = \frac{RB'}{RB + RB'} (-V_{cc}) = \frac{5 \text{ k}\Omega}{25 \text{ k}\Omega} (-12 \text{ V}) = -2.4 \text{ V}$$

This is a most significant determination, and tells us that the voltage divider is trying to make the base *more negative* than the emitter is allowed to go. Since the emitter must try to be nearly the same potential as the base, the base is held to nearly −2 V rather than −2.4 V. The difference can be attributed to the abnormally large base current flowing through RB. This causes the voltage at the base to be more positive than the normal drop across RB when not considering I_B. That is, the true drop across RB is the combined result of current drawn by the biasing resistors and the base current flowing only through RB (and RE eventually).

The circuit, as shown, has biased the transistor far into saturation, and it is completely incapable of amplifying. Without knowing the emitter limits, we could not have easily determined the bias condition of the circuit.

An interesting application of the limits of such a circuit is to use them to help set the proper voltage at the emitter for operation well within the active region. In this example, V_E at ground indicates that the transistor is in cutoff, while V_E at −2 V occurs with the transistor in saturation. Halfway between these limits should be halfway between cutoff and saturation.

The midpoint between 0 and −2 V is −1 V, and by changing the bias resistors so that the base is at −1 V, instead of −2.4 V, the transistor will be biased well within the active region, actually at nearly the exact center of the load line.

One method for finding a new value for RB, assuming that RB' is to remain at 5000 Ω, is to use the fact that the voltage drops across the dividers RB and RB' are directly proportional to the value of the resistors. This, of course, assumes that the base current is small relative to the resistor current. As a proportion:

$$E_{RB'} : RB' :: E_{RB} : RB$$

Solving for RB:

$$RB = \frac{E_{RB} \times RB'}{E_{RB'}} = \frac{11 \times 5000}{1} = 55{,}000 \text{ }\Omega$$

By changing RB to 55 kΩ, the biasing condition is changed.

$$-V_B = \frac{RB'}{RB + RB'} \, (-V_{CC}) = \frac{5 \text{ k}\Omega}{55 \text{ k}\Omega + 5 \text{ k}\Omega} \, (-12) = -1 \text{ V}$$

Since $V_E \cong V_B = -1$ V, the emitter current is set by V_E and RE.

$$I_E = \frac{E_{RE}}{RE} = \frac{1}{2 \text{ k}\Omega} = 0.5 \text{ mA}$$

$I_C \cong I_E$, so the drop across RL is easily determined.

$$E_{RL} = I_C RL = 0.5 \text{ mA} \times 10 \text{ k}\Omega = 5 \text{ V}$$

The collector voltage, referred to ground, is $-V_{CC}$ less the drop across the load resistor.

$$-V_C = -V_{CC} + E_{RL} = -12 + 5 = -7 \text{ V}$$

Also, V_{CE} is equal to the supply voltage less the drop across the load resistor and the emitter resistor.

$$V_{CE} = -V_{CC} + E_{RL} + E_{RE} = -12 + 5 + 1 = -6 \text{ V}$$

and

$$\gamma = \frac{V_{CE}}{V_{CC}} = 0.5$$

Thus it is apparent that the transistor is biased at the center of the load line, since V_{CE} is one-half of V_{CC}. The use of the concept of circuit limits can give us a good idea of where the transistor is biased. From the limits we can also determine the maximum peak-to-peak voltage swing at the collector and, if RE is not bypassed, the maximum emitter swing. Section 6-8 deals with this aspect of circuit behavior.

6-7 TEMPERATURE CHARACTERISTICS

Transistors are fundamentally very sensitive to changes in temperature. For this reason, many of the biasing techniques available for use are seldom seen. It is one of the purposes of any biasing arrangement to help the transistor overcome its inherently poor stability. In Sec. 6-2 several methods of biasing are shown. Those classed as beta-dependent circuits have inherently poor thermal stability, while those classed as beta-independent are generally better in this respect.

The factor that produces this instability is a component of collector current called I_{co}, or I_{cbo}. The magnitude of I_{cbo} is a function of temperature only, and at room temperature is typically less than 5 μA for germanium transistors. However, I_{cbo} doubles in value for every 10°C rise in temperature, and it is conceivable that collector current itself could double because of a large increase in temperature. This, at best, could cause a large shift in the Q point, and at worst could result in destruction of the transistor.

There are several ways to combat this instability. One obvious way would be to operate the transistor in a cold atmosphere. Usually, the refrigeration equipment is costly and bulky, and can be used only in a very large installation. Another way, often used with a power transistor, is to mount it on a "heat sink," so as to carry away the heat generated by the transistor itself. In most cases, proper circuit design can stabilize a circuit so that operation within a reasonable range of temperature is possible.

The bias circuit itself is one means whereby a transistor can be operated satisfactorily at elevated temperatures. Several components in the biasing system are used mainly for their effect upon the temperature stability. Every one of the resistors in the so-called universal circuit, such as Fig. 6-3, has an influence on the temperature stability.

The load resistor RL is limited in value by I_{cbo}. The larger RL is made, the greater the drop across it for a given change in I_{cbo}. Hence values much in excess of 20 kΩ are seldom seen in practical germanium transistor circuits, with 100 kΩ or greater possible with silicon units. The change in the Q point for even a small increase in temperature is great if RL is made excessively large. Usually, values from 1 to 20 kΩ are used.

The emitter resistor RE, sometimes called a *swamping*, or *feedback*, resistor, has a large effect on how the circuit responds to temperature changes. Generally, the larger RE, the better the temperature stability. That this is true is evident from the fact that if I_C increases, so must I_E. Greater I_E means a greater drop across RE, and V_E swings more negative (PNP case). V_E tending to go more negative gives the same result as the base tending to go more positive, and the transistor turns off slightly. This compensates for the original increase in I_C due to a temperature increase.

RB also affects the temperature stability, and when coupled with RB', sets the voltage at the base to a value determined by the resistors themselves. That is, if RB and RB' are made small, the current through them will be large, and the tiny base current from V_{CC} through RB will not affect the voltage at their junction, since the resistive current is much larger. On the other hand, if the resistors are made so large that the resistive current is about the same magnitude as I_B, then an increase in I_B due to a temperature change *can* influence the value of V_B. If V_B goes more negative, the transistor is allowed to turn on harder, and the voltage divider has done little to help. Usually, the current through the divider is made 10 or more times the quiescent base current. A moderate increase in base current represents such a small part of the total current that V_B changes hardly at all.

We thus see that if due consideration is given to all resistors in the bias system, optimum temperature stability can be obtained. It should be pointed out that temperature changes occur slowly; so it makes no difference whether or not RE is bypassed, as far as stability is concerned. Of course, such things as gain, input resistance, etc., will be greatly affected by whether or not RE is bypassed. Chapter 11 deals in detail with the temperature response of several typical circuits.

6-8 LARGE-SIGNAL BEHAVIOR

The typical transistor amplifier shown in Fig. 6-10 can be described in terms of its large-signal response. By this we mean that the collector voltage is made to change over a very large range, in some cases over a range nearly equal to V_{CC}. This is opposed to the small-signal operation, where the collector-voltage change is a very small part of the total applied dc voltage.

In Sec. 6-6 we discussed the limits of a circuit as applied to the

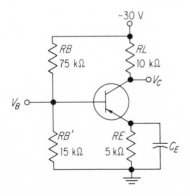

FIG. 6-11. Example of large-signal response.

biasing conditions of the circuit. We can apply the same ideas to determine the maximum possible collector change, and from this the maximum emitter voltage change (if any) and the base voltage change for signal conditions. This results in useful information, since, if a transistor used as a linear amplifier is allowed to either saturate or cut off, severe distortion occurs. By determining the maximum possible collector swing, we can tell at what point this severe distortion will occur.

In the example of Fig. 6-11, the first determination to be made is that of the dc operation point. We must know this, particularly V_{CE}, V_E, and I_E, before proceeding to the ac analysis. The base voltage V_B is set by the voltage divider.

$$-V_B = \frac{RB'}{RB + RB'} (-V_{CC}) = \frac{15 \text{ k}\Omega}{90 \text{ k}\Omega} (-30) = -5 \text{ V}$$

If the base is at -5 V, the emitter must also be at -5 V (neglecting the junction drop); so I_E can now be determined.

$$I_E = \frac{|V_E|}{R_E} = \frac{5}{5 \text{ k}\Omega} = 1 \text{ mA}$$

Since for our purposes $I_C \cong I_E$, the collector current is 1 mA. The drop across RL can now be determined.

$$E_{RL} = I_C \times RL = 1 \text{ mA} \times 10 \text{ k}\Omega = 10$$

Thus

$$-V_C = -30 + 10 = -20 \text{ V}$$

Now V_{CE} is equal to V_{CC}, less the drops across RL and RE; so

$$V_{CE} = -V_{CC} + E_{RL} + E_{RE} = -30 + 10 + 5 = -15 \text{ V}$$

In this instance, the circuit gamma indicates that the bias point is at the exact center of the dc load line.

$$\gamma = \frac{V_{CE}}{V_{CC}} = \frac{15}{30} = 0.5$$

Once the dc conditions are specified, the large-signal response can be determined. Keeping in mind that we are now concerned with signal frequencies, the reactance of C_E is assumed to be zero. The voltage changes to be discussed are therefore assumed to be very fast changes. An ac equivalent can be drawn showing that the emitter is at *signal ground,* even though it is riding above ground by some dc voltage.

Figure 6-12 indicates that the emitter is, for all practical purposes, returned to a dc source of −5 V, this being the purpose of C_E. With

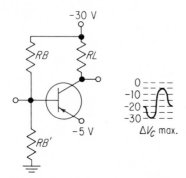

FIG. 6-12.   Collector limits.

the equivalent circuit in mind, the maximum limits of V_C can be calculated.

As always, two conditions serve to set the limits of the circuit. If the transistor is cut off, no collector current can flow, and V_C referred to ground is $-V_{CC}$, or −30 V. This is one limit of V_C. The other limit, then, is with the transistor in saturation. V_C can go no more positive than −5 V, since the emitter is held to this value by C_E. Hence the other limit is −5 V. The collector voltage can therefore range between −30 and −5 V. These limits are called the *large-signal limits* of the circuit.

Note that if the collector were to make this excursion, because of a signal, this would produce a nonsymmetrical output. Quiescently, V_C is at a level of −20 V. It can swing from −20 to −30 V in the negative direction (10-V change), and from −20 to −5 V in the positive direction (15-V change).

To maintain symmetry, the maximum peak-to-peak excursion can be only *twice that of the smaller of the two peaks* illustrated. The maximum symmetrical output excursion, then, is two times a 10-V change, or 20 V peak to peak.

If the base were made to change very slowly, the dc limits of the circuit would be manifest. If the transistor is cut off for a long time, the emitter voltage must fall to ground as C_E discharges. Thus $V_{CE(max)}$ for dc conditions is equal to $-V_{CC}$, while for fast changes it must be something less, in this case 25 V.

Since $V_{CE(max)}$ for signal conditions is less than V_{CC}, the dc load line must be changed for these signal conditions. A new load line is constructed on the curves that is called the *ac, or signal, load line.* Figure 6-13 shows both a dc load line and an ac load line. The dc load

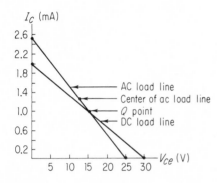

FIG. 6-13.   Dc and ac load lines.

line is constructed between the limits of -30 V and 2 mA. In the case of the ac load line, the limits are $V_{CE(max)} = -25$ V, and $I_{C(max)} = 2.5$ mA. Note that the ac load line passes through the Q point. This is always true, and is typical of circuit action as long as the input signal is symmetrical. (See Chap. 10 for a further discussion of the ac load line.)

Now, with a signal applied, the transistor must operate along the ac load line, as long as the signal frequency is relatively high. Note that the center of the dc load line is not the center of the ac load line. This is why, sometimes, the circuit will be biased at some place other than the center of the dc load line. Often the designer wishes to bias at the center of the ac load line to achieve maximum peak-to-peak output.

We have already determined the maximum symmetrical collector swing as being a 20-V change. It is often of interest to know the maximum input voltage necessary to produce this output. To do this, we must first know the voltage gain A_v of the circuit. The gain is determined by the ratio of RL to all *unbypassed* resistance in the emitter lead. In our example, there is no unbypassed resistance other than the ohmic emitter resistance R_{EE}. However, because we are now concerned with the ac, or signal, conditions, the value of R_{EE} is not appropriate. For signal conditions the value of the *ac resistance* of the emitter must be used. This is called r_e and is determined by dividing the number 0.026 by the emitter current (dc). The number 0.026 is derived from the physics of the transistor. From this relationship the ac resistance of the emitter is seen to be related to the dc emitter current, which is a function of the transistor bias.

$$r_e = \frac{0.026}{0.001} = 26 \ \Omega$$

$$A_v = \frac{RL}{r_e} = \frac{10 \ \text{k}\Omega}{26} = 385$$

The input swing E_I, then, must be E_O/A_v, since $A_v = E_O/E_I$

$$E_I = \frac{E_O}{A_v} = \frac{20}{385} = 52 \ \text{mV}$$

The maximum input excursion necessary to drive the output to 20 V peak to peak is therefore 0.052 V. Usually, a circuit is not driven to its limits since the transistor begins to distort very badly when this is done.

6-9 SMALL-SIGNAL BEHAVIOR

The small-signal behavior of a transistor amplifier can be made to seem quite complex. In advanced studies of transistor circuits the accurate analysis of small-signal conditions requires mathematical techniques that are often beyond the average electronic technician. Fortunately, such rigorous treatment is not necessary for a good understanding of *basic* circuits. By adhering to the principle that a circuit description should be easy to understand and simple to manipulate, very valuable approximations can be made. These, in turn, will allow one to visualize circuit action in a simplified, yet valid, way.

At this time, then, we are concerned with only three small-signal parameters that can easily be determined and verified. These are the voltage gain A_v, the input resistance r_i, and the frequency response F_α and F_β. We shall determine these three values for two very similar circuits, so that the effect of unbypassed resistance can be clearly seen. As we shall see, the circuit with the greater value of unbypassed resistance in the emitter lead has the smaller voltage gain, all else being the same.

First, let us again calculate the voltage gain of the circuit of Fig. 6-11. As previously determined, the value of r_e is 26 Ω, since the emitter current is 1 mA. This is one of the factors that must be known to determine the voltage gain. The other is simply the value of the load resistor. The voltage gain is approximately equal to the value of the load resistor divided by the unbypassed resistance in the emitter lead, including r_e. In this case, there is no physical resistance in the emitter lead; so only r_e can be used.

$$A_v = \frac{RL}{r_e} = \frac{10 \text{ k}\Omega}{26} = 385$$

With a voltage gain of 385, the circuit can amplify any voltage impressed upon its input by this amount. Thus a 1-mV signal at the input will result in a 385-mV signal at the collector.

Now let us calculate the voltage gain of the circuit of Fig. 6-14.

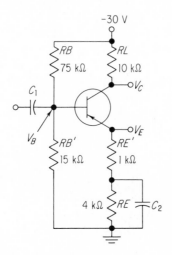

FIG. 6-14. Small-signal example.

The dc condition of the circuit is identical with that of Fig. 6-11, since the total resistance in the emitter lead is the same. However, in this circuit there is an *unbypassed* resistor in the emitter lead. Its value must be included in the determination of A_v.

$$A_v = \frac{RL}{r_e + RE'} = \frac{10 \text{ k}\Omega}{1026} = 9.75 \cong 9.8$$

Note that the large unbypassed emitter resistor greatly reduces the voltage gain of the circuit. However, this effect is not all bad. By reduction of the gain, several other features are improved, notably the bandpass and the linearity. The highest frequency satisfactorily amplified by this circuit is far higher than it would be if the emitter resistor were completely bypassed. Also, the circuit will exhibit far less distortion than it otherwise would. A transistor will, if no negative feedback exists, produce a rather large distortion for anything but very small collector excursions, and the unbypassed emitter resistor greatly reduces this distortion.

The next item to be determined is the input resistance r_i. The total input resistance "seen" by the signal source is a combination of several signal paths. The power supply is at signal ground because of heavy bypassing, and the two resistors, RB and RB', are actually in parallel to the signal current. Thus signal current will flow through each of these resistors, and is lost to the transistor itself.

Another path for signal current is through the base lead of the transistor itself. This is the only part of the total signal current that can affect the transistor. The resistance seen by the source looking from the base toward ground we call R_{ib}. This is equal to $(\beta + 1)$ times all unbypassed resistance in the emitter lead, including r_e. (R_{ib} is the total ac resistance from the base to ground, excluding the base-biasing resistors.)

For the example of Fig. 6-14, R_{ib} is determined as follows:

$$R_{ib} = (\beta + 1)(r_e + R_{E'}) = (99 + 1)(26 + 1000) = 102.6 \text{ k}\Omega$$

Note that this is a very high value for a transistor amplifier, which is usually considered to have a low input resistance. R_{ib} has this high value because of the unbypassed resistance in the emitter lead. Without $R_{E'}$, R_{ib} would be on the order of 2.6 kΩ.

Now we can calculate the total input resistance of the circuit, r_i. This is the combined parallel resistance of RB, RB', and R_{ib}. The easiest way to calculate this is to make the determination in two steps. (The parallel resistance of RB and RB' is RB_{eq}.)

1. $RB_{eq} = \dfrac{RB \times RB'}{RB + RB'} = \dfrac{75 \text{ k}\Omega \times 15 \text{ k}\Omega}{75 \text{ k}\Omega + 15 \text{ k}\Omega} = 12.5 \text{ k}\Omega$

2. $r_i = \dfrac{RB_{eq} \times R_{ib}}{RB_{eq} + R_{ib}} = \dfrac{12.5 \text{ k}\Omega \times 102.6 \text{ k}\Omega}{12.5 \text{ k}\Omega + 102.6 \text{ k}\Omega} = 11 \text{ k}\Omega$

The total input resistance to an ac signal, then, is 11 kΩ for this circuit configuration.

The frequency response of an amplifier is often of interest. Although an accurate determination of the bandpass is very complex, depending as much on construction techniques as on the circuit constants, we can make useful approximations. The transistor manufacturer often specifies the maximum frequency that the product can amplify. This information may be given in one of two ways: The alpha cutoff frequency f_α is the frequency at which the transistor can operate when used in the common-base configuration. The beta cutoff frequency f_β gives the same information for the common-emitter configuration. These figures refer only to the way the transistor itself affects the frequency response, not the way the overall circuit responds to the signal.

Let us assume that a transistor used in a simple *CE* circuit such as Fig. 6-15 has an alpha cutoff frequency f_α of 500 kHz and an ac beta

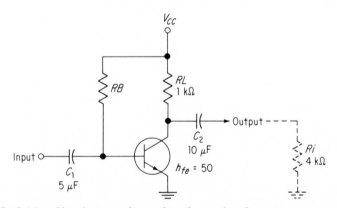

FIG. 6-15. Circuit example used to determine frequency response.

h_{fe} of 50. We must convert the alpha cutoff frequency to beta cutoff frequency f_β, because our circuit is common emitter, not common base. The conversion formulas are

$$f_\alpha = \beta f_\beta \quad \text{and} \quad f_\beta = f_\alpha/\beta$$

In our case

$$f_\beta = \frac{f_\alpha}{\beta} = \frac{500}{50} = 10 \text{ kHz}$$

Thus the transistor in this simple circuit can operate and amplify properly up to 10 kHz in the common-emitter configuration. Its actual maximum frequency may be even less, depending on the circuit stray capacitance, the size of RL, and the transistor capacitance. It should be noted that under some conditions the CE circuit can be made to have very good high-frequency response, perhaps even approaching that of the CB circuit. In Chapter 12 an example is given that relates the actual high-frequency response to the voltage gain of the circuit.

The low-frequency response can be approximated by noting the point at which the frequency of operation causes the reactances in the circuit to affect the gain. In the circuit, for example, both C_1 and C_2 will be found to cause the gain to be reduced at the lower frequencies. The coupling capacitor C_1 will begin to reduce the overall gain at a frequency that causes the reactance to be equal to the total series resistance. Ignoring source resistances, we find that this is roughly where X_{C1} is equal to the input resistance of the transistor, r_{ib}. If transistor collector current is 1 mA, r_{ib} is then:

$$r_{ib} = (\beta + 1)\, r_e = 51 \times 26 = 1326\ \Omega$$

To determine the frequency at which the reactance of C_1 equals the input resistance of the transistor:

$$f_{\text{low }1} = \frac{1}{2\pi R C_1} = \frac{1}{6.28 \times 1326 \times 5\ \mu\text{F}} \cong 24\ \text{Hz}$$

Hence, at about 24 Hz, the reactance of C_1 will begin to reduce the gain of the amplifier.

Also, at some low frequency, the reactance of C_2 begins to reduce the gain. This occurs at a frequency that causes the reactance of C_2 to equal the total resistance presented to it. Assume that the transistor is looking into a load of 4 kΩ.

$$f_{\text{low }2} = \frac{1}{2\pi R C} = \frac{1}{6.28 \times 4000 \times 0.00001} = \frac{1}{0.2512}$$

$$= 3.98 \cong 4\ \text{Hz}$$

The frequency at which both capacitors begin to influence the gain is:

$$f_{\text{low}} = \sqrt{(f_{\text{low }1})^2 + (f_{\text{low }2})^2} \cong 24.33\ \text{Hz}$$

QUESTIONS AND PROBLEMS

6-1 Refer to Fig. 6-4. Change RL to 7.5 kΩ. Determine gamma.

6-2 A transistor is operating with a steady-state V_{CE} of 3 V. V_{CC} is 20 V. What is gamma?

6-3 Refer to Fig. 6-11. Change RL to 7.5 kΩ. Determine gamma. Draw a simplified dc equivalent circuit with all resistances given the proper value.

6-4 Refer to Fig. 6-11. Do not change any values, but remove CE. Determine the minimum and maximum emitter voltage limits. Determine the minimum and maximum collector voltage limits.

6-5 In the circuit of Fig. 6-14, what is the purpose of the capacitor in parallel with RE?

6-6 Refer to the circuit of Fig. 6-10. Determine the large-signal limits of V_B; V_C; V_E; I_E.

6-7 In the circuit of Fig. 6-14, determine R_{ib} if RE' is changed to 2 kΩ and RE is changed to 3 kΩ. (Assume $\beta = 100$.)

6-8 Briefly discuss the reason why the maximum output voltage is often greater in value than the maximum symmetrical output.

6-9 Refer to Fig. 6-10. Determine the signal voltage gain A_v if $\beta = 100$ and $RB = 55$ kΩ.

6-10 Using the circuit of Fig. 6-14, change the value of RE to 2 kΩ. Determine (a) A_v; (b) r_i; (c) maximum symmetrical limits of V_C. The transistor is germanium, and h_{FE} $(\beta) = 100$.

6-11 Refer to Fig. 6-11. The circuit is to be used as shown, but $-V_{CC}$ is to be changed to -12 V, with beta equal to 100. Calculate the following values: V_B, V_C, V_{CE}, γ.

6-12 The circuit values used in Question 6-11 are to be used to determine the large-signal limits of the transistor. Assume the signal frequency is high enough to cause any appreciable capacitive reactance to be essentially zero. Find $V_{CE(max)}$; $V_{CE(min)}$; $V_{C(max)}$; $V_{C(min)}$.

6-13 Refer again to the circuit values of Question 6-11. Briefly discuss why the peak-to-peak output voltage is smaller for this circuit than was the case with the original circuit values in Fig. 6-11.

6-14 The circuit values of Question 6-11 are to be used for a small-signal amplifier. Determine the small-signal voltage gain A_v.

6-15 The circuit values of Question 6-11 are to be used for a small-signal amplifier, but RL is to be changed to 7500 Ω. Find the signal gain A_v.

6-16 A circuit, similar to Fig. 6-11, but with different values, is to be analyzed. The new circuit values are: $RB = 47$ kΩ; $RB' = 4.7$ kΩ; $RE = 1.0$ kΩ; $V_{CC} = 12$ V; $Q_1 = $ NPN; $\beta = 150$; $V_{BE} = 0.7$ V. Determine the quiescent value of collector voltage V_C referred to ground.

Questions 6-17 through 6-20 refer to the circuit just defined.

6-17 Determine the quiescent value of the emitter voltage V_E.

6-18 Determine the quiescent value of base voltage V_B.

6-19 Determine the gamma of this circuit.

6-20 Briefly discuss what effect(s) a gamma of 0.82 will have on the large-signal characteristics of this circuit.

CHAPTER 7
THE COMMON-COLLECTOR CIRCUIT

When a transistor is used in the common-collector configuration, it is usually referred to as an *emitter-follower.* An emitter-follower has many special attributes that make it differ radically from any other kind of circuit. For this reason a rather detailed discussion of the circuit and its application is called for.

7-1 GENERAL DESCRIPTION

The emitter-follower (*CC* circuit) is not a voltage amplifier. It will, however, provide current amplification, and it is therefore very useful. The voltage gain of the emitter-follower is on the order of unity, but the current gain can be 100 or greater. The name of the circuit, the *emitter-follower,* is descriptive of the way it works. Stated simply, the emitter follows the base, and any voltage presented to the base will be copied by the emitter. If this were its only attribute, the circuit would have little use. But because of its ability to amplify current, it finds widespread use in many applications.

The elementary common-collector circuit is shown in Fig. 7-1, and a practical circuit is shown in Fig. 7-2. The input is applied to the

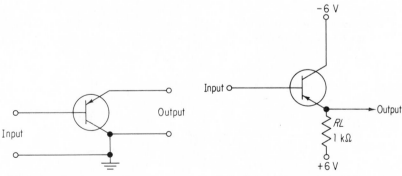

FIG. 7-1. Common-collector configuration.

FIG. 7-2. The emitter-follower.

base, and the output is taken from the emitter. This is an unusual arrangement, and it exhibits equally unusual characteristics.

In order to understand how the emitter-follower functions, one must keep in mind the fact that as long as the emitter-base junction is forward-biased, the base will draw current, and the transistor will be on. To turn off the transistor, the emitter-base junction must be reverse-biased, or at least not biased in the forward direction. In Fig. 7-2, if the base is slightly more negative than the +6-V supply, the transistor is on to some degree.

An important rule to remember is that for the circuit to act as an emitter-follower, it must be in the active region. As long as this is true, the emitter will faithfully follow, or copy, the input applied to the base. That is, if the base is at ground potential, the emitter will be at ground too, except for the small drop across the junction. If the base is lifted to, say, 3 V more positive than ground, the emitter will follow along and will also be at +3 V. As long as the base voltage rides between the total power-supply voltages, any signal appearing at the base will be copied by the emitter, ignoring the junction drop.

The amplification of current is evident when one considers that the emitter current is greater than the base current by the factor beta plus one ($\beta + 1$), as long as the transistor is in the active region. This means that the input circuit requires very little driving current, while the output circuit can deliver a very large current to the load.

7-2 CIRCUIT OPERATION

One way to better understand the emitter-follower is to compare it with circuits that one is already familiar with. Figure 7-3 shows four drawings, starting with an inverter in the *CE* circuit, and progressing toward the emitter-follower. In Fig. 7-1a, the circuit is an inverter with the emitter at ground. If the base voltage is more positive than ground, the emitter-base junction is reverse-biased, and the transistor is turned off. If the base is exactly at ground potential, the junction is neither forward- nor reverse-biased, and again the transistor is turned off. But if the base is made to go somewhat more negative than ground by more than a few millivolts, the transistor will begin to turn on, and some amount of collector current will flow. Because the emitter is firmly tied to ground, the base is "clamped" to ground, and can go to perhaps 0.2 or 0.3 V more negative than ground as a maximum. When this occurs, the transistor is driven to saturation, and the collector voltage falls to ground, very nearly. With the base excursion of something like 0.2 V, the transistor can be driven all the way from cutoff to saturation.

Figure 7-3b is very similar to a, except for the emitter-return circuit. Note that the emitter is returned to +6 V rather than ground.

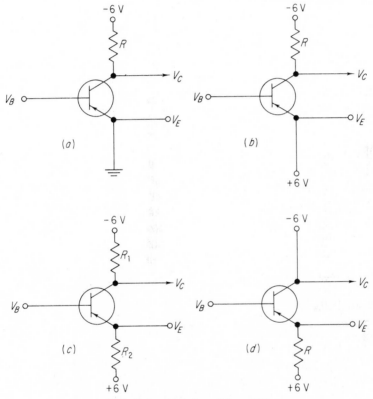

FIG. 7-3.   Relating *CC* to *CE*.

In this case, the base voltage must be more positive than +6 to cut the transistor off by reverse-biasing the junction. But the transistor will begin to conduct if the base becomes slightly more negative than +6 V, say, 5.9 V. Again, the base is clamped to the power supply, and can range only a few tenths of a volt or so and still stay within the confines of the active region. These two circuits are acting in the same way except for the value of emitter voltage.

Figure 7-3c has been modified one step further, and we note the addition of a resistor in the emitter lead. If the base voltage is made more positive than the emitter return, +6 V in this case, the transistor is cut off as we should expect. When the transistor is to be turned on, the base must be driven more negative than the emitter. However, now the emitter is not firmly tied to a power-supply return, and is free to do what the circuit requires of it. If V_B is made to be about 5.7 V, it is 0.3 V more negative than the emitter return. This will cause some amount of base current to flow, and so some emitter current will flow too. With emitter current flowing, a voltage drop occurs across the emitter resistor, and V_E goes more negative. Thus, as the base is caused to go more negative by a signal voltage, the emitter follows because of the increased emitter current that flows. If the base is driven still more negative, more base current flows, which causes still more emitter current to flow, and the emitter voltage goes more negative along with the base. For as far negative as it is allowed to go, the emitter will follow the base, and the two voltages will never be more than perhaps $1/10$ V apart. Now, as the emitter is going in the negative direction because of the increased emitter current, the collector voltage is going in the positive direction, also because of the increased current. When V_E has gone as far as it can, and when V_C has gone as far as it can, the transistor will be in saturation, and the circuit has reached one of its limits. For example, if both resistors are 1 kΩ, the maximum excursion of both V_C and V_E can be determined for the circuit shown by simply considering the transistor to be short-circuited. R_1 and R_2 form a voltage divider that sets the value of V_E and V_C for this condition. The maximum current that can flow is the total voltage divided by the total resistance: 12 V divided by 2 kΩ gives 6 mA. Thus 6 V will drop across R_1, and 6 V will drop across R_2, leaving the junction between the resistors at zero, or ground potential. Because of this, V_E can range between +6 V and ground, while V_C can range between −6 V and ground.

The emitter voltage can be seen to be following the base voltage over part of the range between the two power-supply voltages. By the removal of the resistor in the collector circuit, as shown in Fig. 7-3d, the circuit has become a true emitter-follower, and the emitter will follow the base *over the entire range of power-supply voltages*, in this case, from +6 to −6 V. Now, the transistor is in the active region if the base voltage is between +6 and −6 V. This is true because the base must be made at or above +6 V to cut the transistor off, while to saturate it the base voltage must be −6 V or more. Thus the transistor is in the active region if the base is between the power-supply voltages.

7-3 BIASING THE EMITTER-FOLLOWER

The biasing of the emitter-follower is fairly straightforward. Since the input and output voltages are essentially the same, the dc voltage on the emitter will be the same as that on the base, neglecting the junction drop. The biasing network will therefore set the base, and the emitter will set the voltage level for the quiescent condition. Although it is possible to bias the emitter follower by the beta-dependent method, the variations in the transistors' characteristics make this very undesirable. Normally, the circuits used in practical cases are beta-independent.

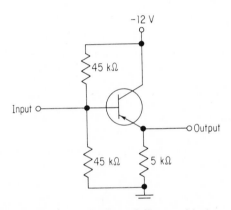

FIG. 7-4.   Emitter-follower biasing.

An example of an emitter-follower is shown in Fig. 7-4. Using this figure as an example, we shall describe the biasing requirements in general terms. To ensure maximum symmetrical output, the base voltage should be halfway between $-V_{CC}$ and ground, or -6 V. The input, and also the output, can swing from -6 V to ground in the one direction, and from -6 to -12 V in the other.

The size of the two base-biasing resistors is determined mainly by the loading requirements of the driving stage. As long as they are equal in value, the voltage at their junction, referred to ground, will be one-half of the total applied voltage, provided only that base current is relatively small.

Closely associated with the biasing requirements of any transistor circuit is the temperature response of the circuit. The emitter-follower has exceptionally good temperature characteristics. Since there is always a rather large resistance in the emitter lead, the dc feedback is 100 percent. Thus one seldom needs to consider the temperature response of the emitter-follower. However, a related point that is sometimes overlooked is the maximum power dissipation of the transistor.

The product of the collector-emitter voltage and the collector current yields the collector dissipation P_C in watts. For instance, suppose a certain transistor is rated at 150 mW maximum. Also, $I_{C(max)}$ is 300 mA. Now, if 300 mA is to be passed by the transistor, what is the maximum collector-emitter voltage V_{CE} that will just allow 150 mW to be dissipated at the collector junction?

$$P_C = I \times E \qquad \text{so } E = \frac{P_C}{I} = \frac{0.150}{0.300} = 0.5 \text{ V}$$

An emitter-follower is often used in a circuit where nearly maximum current is flowing, and since it is always in the active region, there is a danger that the maximum power dissipation will be exceeded. One way of avoiding this excessive power dissipation is to add to the circuit another resistor placed in the collector circuit, as shown in Fig. 7-3c. Now, as current increases because of the signal, the collector voltage falls toward ground, thus lowering the collector-to-emitter voltage as the collector current increases. Thus the product of E and I tends to remain small, and the power dissipated at the collector is kept to a minimum.

7-4 Q-POINT ANALYSIS

We shall want to know several things about the typical emitter-follower when we analyze for the quiescent condition. V_B, V_E, and I_E are rather easily determined. Also, the gamma of the circuit will give us the same kind of information about the emitter-follower that it does for the common-emitter circuit.

As an example, we shall use the circuit of Fig. 7-5. In this circuit, the voltage divider in the base circuit sets the base voltage; so the first step is to find the base voltage V_B, referred to ground. The drop across RB' can be calculated first.

$$E_{RB'} = \frac{RB'}{RB + RB'} \, (V_{total}) = \frac{5 \text{ k}\Omega}{20 \text{ k}\Omega} \times 20 = 5 \text{ V}$$

Now the base voltage can be found.

$$V_B = V_{EE} - E_{RB'} = 10 - 5 = +5 \text{ V}$$

Since V_B is 5 V positive with respect to ground, the emitter is very nearly the same too. Neglecting the drop across the junction, if V_E is 5 V, we can find a value for the emitter current, since this must flow through the emitter resistor.

$$I_E = \frac{E_{RE}}{RE} = \frac{5}{1 \text{ k}\Omega} = 5 \text{ mA}$$

Because I_E and I_C are very nearly the same, the collector current is also 5 mA, for all practical purposes.

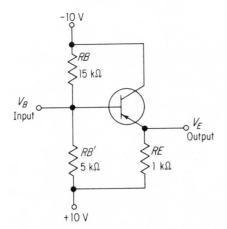

FIG. 7-5. Emitter-follower example.

We can see that the quiescent currents and voltages are quite simple to determine for this circuit.

The concept of gamma for the emitter-follower is no different from that for the common-emitter circuit. The ratio of V_{CE} to V_{CC} conveys the same ability to visualize the point along the load line where the transistor is biased. A gamma of 0.5 indicates the midpoint of the load line, while a gamma of 1 is cutoff, and about 0.05 is saturation. For the circuit of Fig. 7-5, gamma is

$$\frac{V_{CE}}{V_{\text{total}}} = \frac{15}{20} = 0.75$$

The load line for the emitter-follower can be constructed upon the collector curves for the common-emitter configuration. Nor-

mally, we think of the collector curves as applying only to the common-emitter circuit, but they are just as applicable to the other two
configurations. It is useful to alter the procedure slightly when using
these curves for the emitter-follower. After the load line is drawn,
several points along the load line can be plotted that are values of
base voltage, as shown on the load line of Fig. 7-6a. Base voltages of
-1, -3, -5, -7, and -9 V are shown, along with the appropriate
gamma for each position. The load line is drawn for the accompanying schematic.

One limit of the load line is determined by the condition existing
where the transistor is cut off. In this instance, V_{CE} is equal to V_{CC}, or
10 V. The other limit is $I_{C(max)}$, and is determined in the same manner
as for the common-emitter circuit. The maximum collector current
is V_{CC} divided by RE, or 10/1 kΩ, which is 10 mA.

To calculate for the several points along the load line, assume that
the base is made to be -1 V. Assuming that the emitter voltage is the
same, there is 1 V across the emitter resistor; and so 1/1 kΩ is the
amount of emitter current that will flow under these conditions, or
1 mA. Since there is 1 V across the emitter resistor, this leaves 9 V
across the transistor; so V_{CE} is 9 V. This point, then, is plotted at the
junction of the load line, an I_E of 1 mA, and a V_{CE} of 9 V. Note that
gamma is $V_{CE}/V_{CC} = {}^9\!/_{10}$, or 0.9.

Other points on the load line are similarly determined and
plotted. Thus common-collector information is added to the common-
emitter curves, and the curves now relate to this configuration.

If the emitter is not returned to ground, but rather to a separate

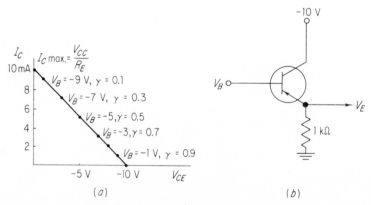

FIG. 7-6.   Emitter-follower gamma.

voltage source V_{EE}, the total difference between the two supplies must be used as the lower extremity of the load line. For instance, if $-V_{CC}$ is -5 V and V_{EE} is $+5$ V, the load line used in Fig. 7-6 will still apply for a load resistor of 1 kΩ.

7-5 SIGNAL RESPONSE

The signal response of the emitter-follower is quite straightforward. The maximum circuit limits have already been discussed. The voltage gain, of course, is unity, for all practical purposes, while the current gain is equal to $\beta + 1$. This leaves only the input resistance r_i of the circuit to determine.

One of the useful attributes of the emitter-follower is the high input resistance that it affords. Because of the large amount of un-bypassed resistance in the emitter lead, R_{ib} is very high. As an example, the circuit of Fig. 7-7 is offered.

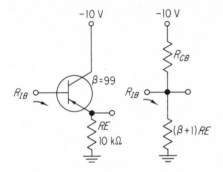

FIG. 7-7. Emitter-follower input resistance.

The ac input resistance of the transistor is $(\beta + 1)$ times greater than the ohmic value of RE and r_e. In the case of the emitter-follower, this can produce a very high value of resistance. (If RE is greater than a few hundred ohms, we can neglect r_e, since it will contribute such a small amount to the total.) In the example, the input resistance is determined as follows ($\beta = 99$):

$$R_{ib} = (\beta + 1)(RE) = 100 \times 10 \text{ kΩ} = 1 \text{ MΩ}$$

The loading on the signal source, then, is only that of a 1-MΩ resistance, and not a much lower value, as one would expect with a transistor circuit.

7-6 AN EQUIVALENT CIRCUIT

A simplified equivalent circuit of the emitter-follower can be a great
help in visualizing the reasons behind the major characteristics of
the circuit: high input resistance and low output resistance. Simply
saying that a circuit has a high input resistance and a low output
resistance does not prove the fact. The dc equivalent circuit of Fig.
7-8 can, however, help to prove the "why" of these circuit characteristics. No biasing arrangements are shown because this is not our
present concern. Since the two major circuit characteristics are the

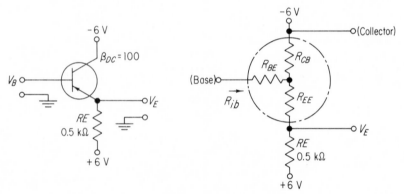

FIG. 7-8.   Dc equivalent circuit.

high input resistance and the low output resistance, these will be
explained in detail. In Sec. 7-5, we briefly discussed the input resistance. Now we want to see the reason for this.

The basic circuit is shown along with the equivalent circuit that
we shall use to help explain the input-output characteristics. The
transistor itself is shown enclosed in the dotted line. R_{CB}, R_{BE}, and r_e
are the transistor parameters, while RE is, of course, part of the external circuit.

We define the R_{ib} of the transistor as the total input resistance,
base to ground, of the transistor. This can conveniently be thought
of as defining either the dc or signal state of the circuit, with either
the dc or signal value of beta used, and either R_{EE} or r_e used, as appropriate. The total dc value of R_{ib} for this configuration is a combination of several factors, RE having the greatest effect. The equivalent
base resistance R_{BE} is numerically equal to $(\beta + 1)$ times R_{EE}. As an
example, assume that beta is 100 and that R_{EE} is 8 Ω.

$$R_{BE} = (\beta + 1)(R_{EE}) = 101 \times 8 = 808 \ \Omega$$

Viewed from the base, the emitter resistor also appears to be $(\beta + 1)$ times greater than its actual value.

$$RE_{(equiv)} = (\beta + 1)(500) = 101 \times 500 = 50,500 \ \Omega$$

The total input resistance is the combination of these two figures, although it can be seen that the value of R_{BE} is so small that it could be neglected. However, if it must be considered, the following expression will determine the total dc R_{ib} for this circuit:

$$R_{ib} = (\beta + 1)(R_{EE} + RE) = 101 \times 508 = 51,308 \ \Omega$$

Since RE is usually no better than a ± 5 percent resistor, the inclusion of R_{BE} is not important in this case. If R_{EE} is of a reasonable size compared with RE, it would have to be considered in the interest of accuracy. The total dc input resistance of this circuit, then, is about 51 kΩ.

For signal conditions a source will be driving this circuit, and it must supply current to a slightly different value of resistance.

The value of R_{ib} to a signal is only slightly different from the dc values in this instance. The following calculation clarifies this point:

$$R_{ib(ac)} = (\beta + 1)(r_e + RE)$$

$$\text{where } r_e \cong \frac{26}{I_E \ (\text{mA})}, \text{ or } \frac{0.026}{I_E \ (\text{A})}$$

$$\beta = h_{fe}, \text{ or } \beta_{ac}$$

From a set of output curves for a transistor having a dc beta of 100 (for instance, Fig. 5-22), h_{fe} might be derived as 110. The input resistance to a signal is then easily found. (Assume $I_E = 13$ mA.)

$$R_{ib(ac)} = (\beta + 1)(r_e + RE) = 111\left(\frac{26}{13} + 500\right)$$

$$= 111 \times 502 = 55.7 \text{ k}\Omega$$

Note that this is not significantly different from the dc condition, as is usually the case with no external loading.

If, for example, a 1-V peak-to-peak signal is impressed upon the input, the base current will be on the order of 17 μA, also peak to peak. Hence the loading on the signal source is very small.

This discussion explains why the emitter-follower is used so extensively in cases where the loading on a signal source must be minimized. By proper design, the input resistance of a typical emitter-follower can be made to be several megohms if necessary.

The other important feature of the emitter-follower is its low output resistance. To approximate this, all possible shunt paths for signal current must be investigated. The ultimate load looking back into the emitter sees three major paths. One of these, of course, is the resistor RE. Another path is up through the transistor and through R_{CB} and out to signal ground. The third path — and this is not as obvious as the other two — is through the transistor and out the *base* lead to signal ground through any resistance in the base lead. This third path is usually, with normal-sized components, the most prominent one. Any resistance in the base lead appears, when viewed from the emitter, to be less than its actual value by the factor $(1 - \alpha)$. In the usual circuit this is a much lower value than either R_{CB} or RE. As an example, if the total shunt resistance in the base lead of an emitter-follower is 1000 Ω, and if alpha is 0.98, the output resistance is determined as follows:

$$r_o \cong (1 - \alpha)RB = 0.02 \times 1000 = 20 \ \Omega$$

In most circuits this would be much lower than any normal value of RE. Thus the output resistance would be very nearly equal to 20 Ω in this instance.

The emitter-follower is often likened to a transformer because of this ability to change a high impedance, or resistance, to a lower one. Thus the current flowing in the input circuit is very small, whereas the current flowing in the output circuit can be relatively large.

With germanium transistors, the voltage from base to emitter is always on the order of 0.1 V, as long as the transistor is in the active region (not including power transistors). Any voltage that appears on the base, then, will also appear at the emitter lead, changed only by the small drop across the junction. We can easily verify that a small change in the base-circuit current will result in a much larger current change in the emitter circuit, by looking at an example.

Still referring to Fig. 7-8, assume that the quiescent base voltage is 0 V. If a 1-V signal is impressed upon the base, and if the polarity is such as to cause a positive excursion, the current flow in the base circuit that is caused by the signal is E_{sig}/R_{ib}.

$$I_{\text{sig}} = \frac{E_{\text{sig}}}{R_{ib}} = \frac{1}{56 \text{ k}\Omega} = 17 \text{ }\mu\text{A}$$

That is, the base-current *change* is 17 μA, and in this case, since this transistor is a PNP, this is a decrease of 17 μA. Therefore, if the base current changes by this amount, the emitter current must change by $(\beta + 1) \times 17$ μA. This should be 111×17 μA, or $\cong 2$ mA. Since the quiescent emitter current is 12 mA, the emitter current must decrease by 2 mA, which would be a new value of 10 mA.

Because the base-to-emitter voltage is practically constant for small changes, the emitter voltage must be 1 V more positive than it previously was. If we assume that the original value of V_E was 0 V, then, with the signal impressed, it must be $+1$ V. Let us see if the current through RE is the same as the current arrived at above. The voltage across RE is now about 5 V; therefore the current through it must be $E_{RE}/RE = 5/500 = 10$ mA, which is the same value as above.

7-7 APPLICATION

The characteristics of the emitter-follower make it especially suitable for a number of applications. It is particularly useful as an impedance transformer, since it has inherently high input impedance and low output impedance, as we have just seen. The high input impedance, or resistance, is often used in preamplifiers that are operated from a high-impedance source, such as certain microphones, phonograph pickups, etc. A typical example of such a circuit is shown in Fig. 7-9.

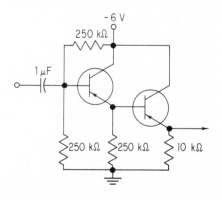

FIG. 7-9. Stacked emitter-follower.

This is a so-called stacked emitter-follower. The input stage provides fairly high input resistance, while the output stage provides the low output resistance. No voltage gain is afforded, but the current amplification is quite large. The output of this preamplifier would normally be fed into the input of a common-emitter amplifier to give the required voltage gain.

The base of the input stage is obviously biased at about −3 V; so its emitter is also at nearly −3 V. Since this is a reasonably stable voltage, there is no need for a second voltage divider in the base of the second transistor. Thus the two stages can be direct-coupled, as shown. The input resistance of the first stage is about the parallel resistance of the two base resistors, or 125 kΩ. The R_{ib} of the transis-

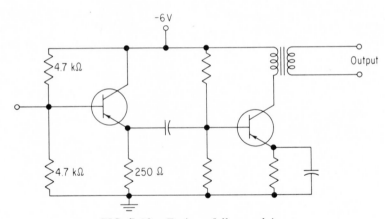

FIG. 7-10.   Emitter-follower driver.

tor is so high that it does not significantly alter the total input resistance. The output resistance of the second transistor is on the order of 1.25 kΩ.

A second example, shown in Fig. 7-10, utilizes the very low output resistance to drive the base of a power amplifier. Such an arrangement requires a large current capability, since the output transistor requires a large amount of base current to deliver maximum power output. The signal beta of a power transistor is typically low, and so the base-current drive must be correspondingly large.

Emitter-followers are not encountered as often in linear circuits as they are when working with pulse circuits. Although it is not the purpose of this discussion to cover pulse circuits, a typical example of emitter-followers used in a pulse application is in order.

The typical emitter-follower in this case is used when the signal is to be sent to several different loads, and the original source of this signal, usually an inverter amplifier, cannot deliver the amount of current needed. Figure 7-11 shows an emitter-follower with several loads, and it is its function to ensure that the signal across each load is the same as its input.

Because of the ability to deliver large currents at low impedance levels, it can ensure that the voltage across each load is a full 6 V, in the example. To maintain the requirement that it must be operated always in the active region, the power-supply levels must be greater than the signal levels. Thus, in the example, the signal must never go to either −18 or to +12 V. (In special cases it may be desirable for the

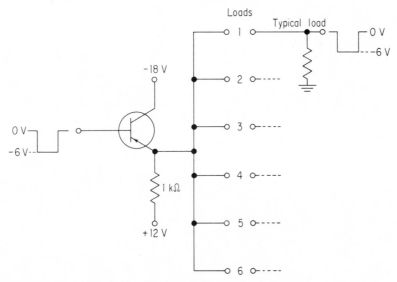

FIG. 7-11.   Emitter-follower for pulse work.

emitter-follower to be driven into saturation; so we must not say that this is never done.) We can see that the operation of the emitter-follower is not much different whether used in a linear circuit or a pulse circuit.

QUESTIONS AND PROBLEMS

7-1   True or false: A common-collector circuit is one that has an output at the collector.

7-2 True or false: The emitter-follower is useful because of the large voltage gain.

7-3 True or false: The emitter-follower is useful in cases where it is necessary to transform impedance levels.

7-4 True or false: As a general rule, an emitter-follower exhibits very good temperature characteristics.

7-5 Select the correct answers. Refer to the circuit shown below.

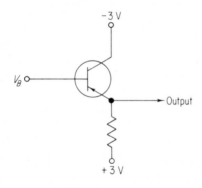

(a) If $V_B = +2$ V, the output is most nearly
    (1) −3 V   (2) −2 V
    (3) +2 V   (4) +3 V
    (5) 0 V
(b) If $V_B = +4$ V, the output is most nearly
    (1) −3 V   (2) −2 V
    (3) +2 V   (4) +3 V
    (5) 0 V   (6) +4 V
(c) If $V_B = -4$ V, the output is most nearly
    (1) −3 V   (2) −2 V
    (3) +2 V   (4) +3 V
    (5) 0 V   (6) −4 V

7-6 Using the circuit of Fig. 7-5, but with RB changed to 10 kΩ, determine the base voltage V_B, at the Q point, and I_E. What is the gamma of this circuit?

7-7 Refer to Fig. 7-7. Change RE to 4.7 kΩ. Calculate the new value of R_{ib}. Now change beta from 99 to 49 and calculate the new value of R_{ib}. Discuss the relation between beta and R_{ib}, all else remaining the same.

7-8 Refer to Fig. 7-9. Assume that the input-transistor beta is 50, and change RE from 250 to 2.5 kΩ. Determine the R_{ib} of the input transistor; determine r_i.

7-9 Refer to Fig. 7-10. What is the r_i of the driver stage? Assume $\beta = 100$.

7-10 Select the correct answers. In the circuit shown

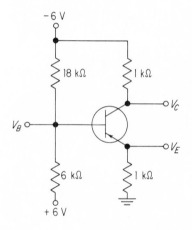

(a) V_B is most nearly
    (1) −6 V   (2) +6 V
    (3) −2 V   (4) +2 V
    (5) −3 V   (6) +3 V
(b) V_E is most nearly
    (1) −6 V   (2) +6 V
    (3) −2 V   (4) +2 V
    (5) +3 V   (6) 0.0 V
(c) V_C is most nearly
    (1) −6 V   (2) +6 V
    (3) −2 V   (4) +2 V
    (5) −3 V   (6) +3 V

7-11 Select the correct answer. The emitter-follower can be partially described by its voltage gain, which is

(a) large   (b) small
(c) > 1   (d) < 1

7-12 Select the correct answer. An attribute of the emitter-follower
is
(a) low-input, high-output resistance
(b) high input, low output resistance
(c) high voltage gain, low current gain
(d) output signal voltage inverted

7-13 Select the correct answer. The emitter-follower is so called
because its emitter follows
(a) the collector   (c) the input signal
(b) the power supply   (d) the bias voltage

7-14 In the circuit shown, determine the voltage referred to ground
at the emitter V_E. Determine the voltage E_{RE} at the emitter
referred to +10 V.

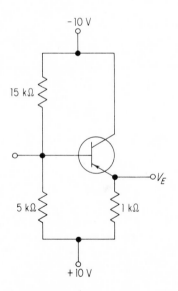

7-15 Briefly discuss why a common-collector circuit is so called.

7-16 In the circuit shown, determine the dc output voltage at X.
Assume negligible V_{BE}.

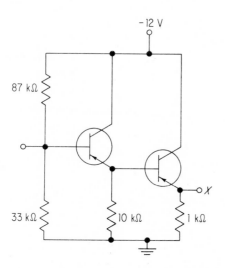

7-17   In the circuit shown, determine gamma.

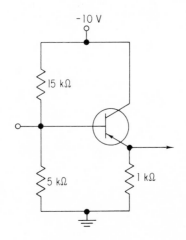

7-18   Refer to Fig. 7-5. This circuit is to be used as shown except for RB' which is to be changed from 5 kΩ to 10 kΩ. Determine the base voltage V_B referred to ground (assume $\beta = 100$).

7-19   Using the altered circuit values of Question 7-18, determine the emitter current I_E.

7-20   Using the circuit values of Question 7-18, determine the circuit gamma.

7-21 Using the circuit values of Question 7-18, determine the circuit power gain A_P.

7-22 Again using the same circuit values, determine the total input resistance (or impedance) r_i.

7-23 Determine the R_{ib} of the foregoing circuit.

7-24 Determine the output resistance R_o of the above circuit.

7-25 It is necessary to change the circuit values of Question 7-18 so that the peak value of output current rises to 30 mA when the transistor is turned fully on. What is the *new* value of *RE*?

7-26 Refer to the circuit values of Question 7-18. Determine the maximum peak-to-peak output voltage, if the circuit is allowed to distort slightly.

7-27 Refer to Fig. 7-5. Determine the new value for *RB'* so as to allow maximum symmetrical output voltage.

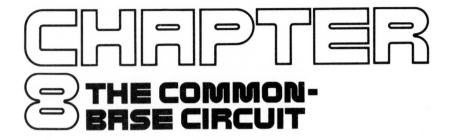

CHAPTER 8 THE COMMON-BASE CIRCUIT

8-1 GENERAL DESCRIPTION

The third kind of circuit in which a transistor can be placed is the common-base circuit. The fundamental circuit configuration is shown in Fig. 8-1, along with a practical example. In this circuit, the input is applied to the emitter, while the output is taken from the collector. The base, then, is at signal ground. As a signal current is applied to the emitter, the collector current is made to vary, and an output variation appears at the collector.

The signal response of this circuit is considerably different from either the *CE* or the *CC* configuration. One of the differences is the lack of current gain. The signal current that must be supplied by the signal source must be as great as (in fact, slightly greater than) the signal current that is caused to flow in the collector circuit.

In the common-base circuit, the emitter-current change—and therefore the collector-current change—must come from the driving source. The current gain of the circuit is the alpha of the transistor;

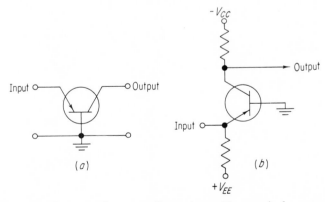

FIG. 8-1. (*a*) Common-base configuration; (*b*) practical common-base circuit.

so I_C is always less than I_E. The voltage gain, however, can be as large as, or larger than, that in the common-emitter circuit.

Looking into the input of this circuit, the signal source "sees" nearly a short circuit. That is, the input resistance of this circuit is very low, typically, 25 Ω or so. The driving source, then, must deliver a very large current from a very low impedance. For this reason the circuit must be driven with either a transformer or an emitter-follower to provide the low-impedance high-current source.

In a practical case this is not always desirable, and the circuit shown in Fig. 8-2 is often used to overcome this difficulty. Here the input resistance is increased by the value of R_S, the coupling resistor. This allows a measurable voltage swing at the input and an easily

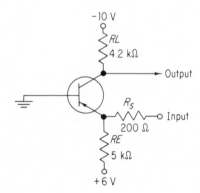

FIG. 8-2.   Analysis example.

calculated change in current through the transistor. Also, the signal source can be a more reasonable circuit, having to deliver only a moderate voltage and current.

As usual, in order to gain something extra from a circuit, something must be sacrificed. In this case the voltage gain is sharply reduced, but the increased input resistance of the circuit is often worth the compromise.

8-2 Q-POINT ANALYSIS

The circuit of Fig. 8-2 is biased by the same method mentioned in Sec. 6-2, shown there as circuit *f*. The two resistors *RL* and *RE* determine the bias point, since there are no base-biasing resistors. To analyze this circuit for its *Q* point, we must first determine the emitter current. With no signal applied, the base is at ground potential, and the base-emitter junction is forward-biased. This is true because the base is returned to a point that is more negative than V_{EE} by 6 V. Thus the emitter junction is forward-biased, and base current must be flowing. For a germanium transistor, the drop across the base-emitter junction seldom exceeds 0.1 V. So if the base is at ground, the emitter must also be very nearly at ground. Assuming V_E to be at ground will allow us to determine the current through the emitter resistor, which must be the emitter current.

$$I_E = \frac{E_{RE}}{RE} = \frac{6}{5 \text{ k}\Omega} = 1.2 \text{ mA}$$

Now, since $I_C \cong I_E$, the drop across RL can be calculated.

$$E_{RL} = I_C \times RL = 1.2 \text{ mA} \times 4.2 \text{ k}\Omega = 5 \text{ V}$$

The quiescent value of V_C is equal to the value of V_{CC} less the drop across the load resistor.

$$-V_C = -V_{CC} + E_{RL} = -10 + 5 = -5 \text{ V}$$

To find gamma for this circuit, it must first be appreciated that the emitter is firmly clamped to ground by the base, and V_{EE} does not affect the limits of the collector. Thus, for this circuit,

$$\gamma = \frac{V_{CE}}{V_{CC}} = \frac{5}{10} = 0.5$$

8-3 DC EQUIVALENT CIRCUIT

The dc equivalent circuit reveals much about the circuit itself. The load resistor and R_{CB} form a simple voltage divider between $-V_{CC}$ and ground. The collector current is adjusted to produce about one-half of V_{CC} across RL and one-half across R_{CB} for normal biasing. The value of R_E is set to provide the proper collector current for these conditions, assuming that gamma is to be 0.5 for the quiescent state.

The base is firmly tied to ground, and the emitter therefore cannot vary away from ground by much more than 0.1 V as a signal is applied. Quiescently, both base and emitter can be considered to be at ground potential; so emitter current, and therefore collector current, can be easily determined. As an example, let us determine the quiescent values for the equivalent circuit of Fig. 8-3. This is the same circuit as shown in Fig. 8-2, but with different values.

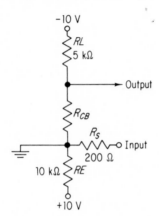

FIG. 8-3.   Dc equivalent circuit of the common-base amplifier.

Since the emitter of the transistor is essentially at gound, the emitter current is determined by V_{EE} and RE.

$$I_E = \frac{10}{10,000} = 1 \text{ mA}$$

Since I_E and I_C are nearly the same, we shall make little error if we assume that I_C is also 1 mA. The drop across RL can now be determined.

$$E_{RL} = I_C \times RL = 1 \text{ mA} \times 5 \text{ k}\Omega = 5 \text{ V}$$

The collector voltage, referred to ground, is simply $-V_{CC} + E_{RL}$.

$$-V_C = -V_{CC} + E_{RL} = -10 + 5 = -5 \text{ V}$$

The value of R_{CB} is a function of V_{CE} and I_C. In this circuit, V_{CE} is equal to V_C, because the base clamps the emitter very nearly to ground.

$$R_{CB} = \frac{V_{CE}}{I_C} = \frac{5}{1 \text{ mA}} = 5 \text{ k}\Omega$$

Gamma, the point along the dc load line where the transistor is biased, is

$$\frac{V_{CE}}{V_{CC}} = \frac{5}{10} = 0.5$$

8-4 CIRCUIT LIMITS—LARGE SIGNAL

The limits of the circuit of Fig. 8-3 are quite straightforward. If the collector rises to $-V_{CC}$, the transistor is in cutoff. If the transistor is in saturation, only RL and V_{CC} determine the maximum collector current, and V_{CE} is nearly zero. Thus the collector voltage can swing from -10 V to ground. Since it is biased so that I_C is quiescently at -5 V, it can swing an equal amount either side of the Q point. It is therefore biased for maximum peak-to-peak symmetrical output (with no external load).

8-5 SMALL-SIGNAL RESPONSE

The parameters of the common-base circuit that interest us are the input resistance r_i, the signal gain A_v, and the frequency response. The input resistance of the circuit of Fig. 8-3 is very easy to determine. For all practical purposes, r_i is 200 Ω. In other words, RS is the total input resistance.

If there were no RS, the input resistance would be very nearly equal to r_e, in this case about 26 Ω. This is so nearly a short circuit that it is evident why RS is used.

The voltage gain of this circuit is determined somewhat differently than before. Usually, the relation RL/RE will yield correct results, but in this case RS is used in place of RE. In the example the voltage gain is

$$A_v = \frac{RL}{RS} = \frac{5 \text{ k}\Omega}{200} = 25$$

This assumes that r_e is much smaller than RS. If not, the denominator becomes $RS + r_e$.

Thus a 10-mV signal applied to the input will result in a 250-mV signal at the collector.

Finally, the frequency response of the circuit can be approximated. The low-frequency response of the circuit of Fig. 8-3 extends to dc, since there is no coupling capacitance. If the transistor manufacturer gave us the beta cutoff frequency for the transistor as 10,000 Hz, we could calculate for the maximum high frequency possible. If beta is 100, then

$$f_\alpha = f_\beta \times \beta = 10 \text{ kHz} \times 100 = 1 \text{ MHz}$$

Thus the maximum possible high frequency (3 dB) for this circuit is 1 MHz. As always, the upper limit is less than this figure because of the circuit stray and distributed capacitance.

8-6 CIRCUIT EXAMPLE

To see how a signal affects the common-base circuit using RS, we shall describe the circuit of Fig. 8-4. Note that this is a similar circuit, except for the different biasing scheme. We shall not describe the derivation of the dc response, since we have done this many times already. It might be instructive for readers to calculate these values for themselves.

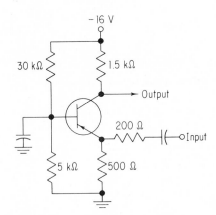

FIG. 8-4. Example for signal conditions.

Quiescently, the emitter is resting at −2.3 V, and about 4.6 mA of collector current is flowing. When a signal appears, the input side of RS is driven, say, negative. At this time there is a voltage across RS equal to the signal voltage. This tends to drive the emitter slightly negative, perhaps only a few millivolts. Driving the emitter negative has the same effect as driving the base positive. Collector current must therefore decrease, and $-V_C$ rises toward $-V_{CC}$ (PNP case).

As the input swings positive, the transistor is turned on harder, and more collector current flows. Again, driving the emitter more positive has the same effect as driving the base more negative. With greater collector current, the collector voltage $-V_C$ falls toward ground. It is evident that there is no phase inversion, since the output follows the input as far as the polarity is concerned.

A simplified ac analysis will reveal the mechanism of circuit operation. Assume a 0.25-V peak-to-peak input signal, with each peak 0.125 V above and below the base line. As the input goes to +0.125 V, the voltage across RS is 0.125 V, with current flow I_s as shown in Fig. 8-5a. The amount of peak current is very nearly equal to

$$I_s = \frac{E_{RS}}{RS} = \frac{0.125}{200} = 0.62 \text{ mA}$$

(This calculation depends upon V_E being perfectly constant, which is not quite the case. However, the error is so slight that it can be safely ignored.)

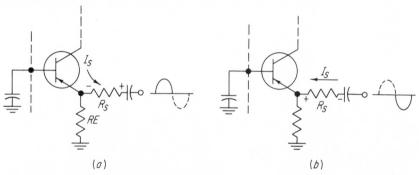

(a) (b)

FIG. 8-5. (a) Signal current, positive input; (b) signal current, negative input.

Now, V_E is essentially constant, and because Ohm's law must be satisfied, the current through R_E must remain constant. Hence the current through RS cannot in any way be contributed by RE. The collector current, then, must increase by 0.62 mA to accommodate this current path, as shown in Fig. 8-5a. Actually, the emitter is driven very slightly positive, and I_B increases slightly, so beta times I_B must flow in the collector and out through RS. The collector voltage falls toward ground by an amount determined by RL and the change in I_C. The *change* in collector voltage, $\Delta|V_C|$ is therefore

$$\Delta V_C = \Delta I_C \times RL = 0.62 \text{ mA} \times 1.5 \text{ k}\Omega = 0.93 \text{ V}$$

The collector voltage, then, falls toward ground by nearly 1 V.

When the input swings negative, current now flows through RS in the opposite direction, and collector current must decrease. The amount of decrease is again equal to the *change* in collector current that occurs in RL. This condition is shown in Fig. 8-5b.

$$\Delta V_C = \Delta I_C \times RL = 0.62 \text{ mA} \times 1.5 \text{ k}\Omega = 0.93 \text{ V}$$

Now the collector voltage rises toward $-V_{CC}$ by nearly 1 V.

The total change in output voltage is 1.86 V, while the input is 0.25 V. The output change e_o, divided by the input change e_i, is the amount that the transistor has amplified the signal.

$$A_v = \frac{e_o}{e_i} = \frac{1.86}{0.25} = 7.5$$

An alternative way to calculate the voltage gain is to use the RL/RS ratio.

$$A_v = \frac{RL}{RS} = \frac{1.5 \text{ k}\Omega}{200} = 7.5$$

In this example, the gain is not large, but could be made larger by reducing the size of RS and/or increasing RL. However, this would require more driving power from the source. So in many cases the gain must be sacrificed in favor of reasonable input requirements.

The maximum collector swing possible is determined by the ac voltage limits, one of which is $-V_{CC}$. The other limit is the quiescent value of V_E, for the collector can fall only as far toward ground as V_E, and only then if the transistor is in saturation. Thus the maximum collector swing is $-16 + 2.3$, or 13.7 V. However, since the transistor is biased slightly off center, the maximum collector swing is slightly less than this. The collector voltage, referred to ground, is quiescently -9.1 V. It can swing from -9.1 to -16 V, a 6.9-V excursion. With a symmetrical input the collector can swing only as far positive as it can swing negative. The true symmetrical collector change, then, is twice the distance from the Q point to the nearest limit. In this case, twice 6.9 is 13.8, and this is the maximum collector voltage swing for a symmetrical output.

The maximum input change is simply the change in V_C/A_v.

$$e_i = \frac{\Delta V_C}{A_v} = \frac{13}{7.5} = 1.84 \text{ V, peak to peak}$$

A larger input signal would drive the collector voltage to one of the limits.

In summary, then, we can say that the common-base circuit has excellent high-frequency response, good temperature stability, and high voltage gain. Its disadvantages are a very low input impedance and the lack of current gain.

QUESTIONS AND PROBLEMS

8-1 Select the correct answer. The common-base circuit has the attribute that
(a) the input resistance is very high
(b) the voltage gain is less than 1
(c) the current gain is less than 1
(d) the output resistance is very low

8-2 Select the correct answer. One description of the CB circuit tells us that
(a) the input resistance is very low
(b) the frequency response is poor
(c) the output voltage is inverted
(d) the output current is beta times larger than the input

8-3 True or false: A CB amplifier is often driven by an emitter-follower to ensure large input-current drive.

8-4 Briefly discuss why a common-base circuit is so called.

8-5 Refer to Fig. 8-4. Change the collector load resistor to 2.5 kΩ. What is V_C, referred to ground?

8-6 With the collector load resistor of Fig. 8-4 changed to 2.5 kΩ, what change, if any, occurs to V_E?

8-7 True or false: A common-base circuit has the characteristic of very low input impedance, or resistance.

8-8 True or false: The current gain of a typical common-base amplifier is very high.

8-9 Refer to Fig. 8-2. Change the value of the emitter resistor to 2500 Ω. When $\beta = 100$, what are the new values of V_C; V_B; V_E; I_C?

8-10 Refer to Fig. 8-1. Discuss the reason why the dc input resistance is very low, or essentially zero.

8-11 Refer to Fig. 8-4. Change RL from 1500 to 2500 Ω and RE from 500 to 1000 Ω, but all else remains the same as illustrated. Beta is 135 and V_{BE} is 0.2 V. Determine V_B, V_E, and V_C.

8-12 Refer to the circuit values of Question 8-11. Determine the value of gamma.

8-13 Refer to the circuit values for Question 8-11. Determine the midfrequency voltage gain A_v.

8-14 Refer to the circuit values for Question 8-11. Determine a new value for RS so as to cause the midfrequency gain A_v to be 25.

8-15 Refer to the circuit values for Question 8-11. Determine a new value for RS so as to cause the midfrequency gain A_v to be 1.

8-16 Refer to the circuit values stated in Question 8-11 (no other changes). Determine the power gain (beta is 20).

8-17 Using the circuit values stated in Question 8-11, determine the output voltage e_o if the input voltage e_i is 85 μV peak to peak (at midfrequency).

8-18 Using the circuit values stated in Question 8-11, find the proper value of base bypass capacitor such that at 50 Hz its reactance X_C is $0.1 \times RS$.

8-19 Again using the values in Question 8-11, find the value of the coupling capacitor such that at 100 Hz the reactance X_C is $0.25 \times RS$.

8-20 Using the circuit values of Question 8-11, find the listed large-signal, midband limits (assume $V_{CE\ sat} = 0$ V): $V_{C(max)}$ and $V_{C(min)}$.

CHAPTER 9
DC ANALYSIS METHOD

The first determination to be made when analyzing any transistor circuit, as has been mentioned before, is to calculate for the quiescent values. We have done this for the simpler circuits in a straightforward way. For more complex circuits, however, the simplified method will not work, as we shall see. A search for a better way to analyze for the quiescent values of a circuit led to the equivalent resistance circuit analysis (ERCA) method. This is at least as simple as any other method, yet will allow even the more complex circuits to be easily analyzed.

Our present purpose, then, is to use a more advanced concept of the equivalent resistance R_{CB} to analyze for the dc conditions of a circuit. Then, in a later chapter, we shall use this, plus all our previous methods, to analyze for all three responses: dc, large signal, and small signal. We shall now investigate in greater detail the seven fundamental methods of biasing a transistor circuit. Nearly any linear-amplifier circuit encountered will use one of these circuits with only minor variations.

We have previously discussed the dc equivalent circuit of a transistor, using R_{CB} as the transistor equivalent. To some, such a gross simplification is little short of ridiculous. Because of this reaction, either they discard a basically valuable tool or they apply this principle without realizing it. As an aid in visualizing the relationships in a circuit, the concept of R_{CB} as we have used it thus far is quite valuable enough. But it has an even greater use: that of completely specifying the dc, and often the large-signal responses of the circuit. Even more difficult circuits, with multiple dc feedback, can easily be solved for these quantities.

In Chap. 6 we divided all biasing circuits into seven basic configurations, and these we divided into two groups. The first of these groups we called the beta-dependent circuits, and the second group, the beta-independent circuits. We shall continue to use this grouping in this chapter. So that we may learn the basic ideas in a sensible way, we shall first apply these ideas to the simple circuits and then proceed toward the more complex circuits. The method is more useful for the complex circuits, but to fully understand the principles, we shall analyze the circuits one by one.

Fundamentally, the basic idea is this: We know the value of all physical constants in any circuit when we begin to analyze it. We know the resistor values, the power-supply output, etc. The transistor itself is the only unknown. From the standpoint of the dc values in the circuit, we can replace the transistor with the proper value of R_{CB}. Thus R_{CB} is the only unknown when trying to solve for the dc values in a circuit. If we have a way of easily solving for R_{CB}, we can determine the remaining values by the use of nothing more complex than Ohm's law. The ERCA equations will allow us to do just that. By giving R_{CB} a value, all dc currents and voltages can be determined by the same methods that are used in a simple series-parallel resistor network!

BETA-DEPENDENT CIRCUITS

9-1   CIRCUIT a—FIXED-BIAS CIRCUIT

When a circuit is to be analyzed for its quiescent conditions, the first step is to draw the dc equivalent circuit as we did in Chap. 6. R_{CB} is substituted for the transistor itself, and this represents the reverse-biased collector junction. Figure 9-1a shows a simple circuit along with its dc equivalent circuit.

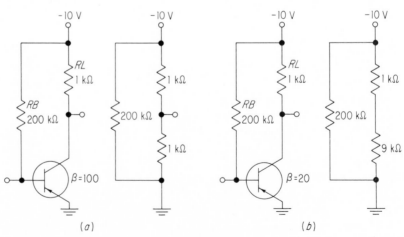

FIG. 9-1.   (a) Fixed-bias circuit, $\beta = 100$; (b) fixed-bias circuit, $\beta = 20$.

There is a fundamental relationship in this kind of circuit that is not clearly evident. We know that the current through the load resistor and the transistor is beta times greater than the current through the base resistor. We also know that the voltage across each parallel branch is essentially the same. Since this is true, it is also true, although it is not evident, that *the resistance in each branch must be related.* The factor by which these two branches are related must be *beta.* Some thought will prove that this must be so.

The current in the collector of a transistor is specified as beta times the base current. Thus it does not make any difference how much resistance is placed in the collector lead of the transistor as far as *collector current* is concerned. If *RL* is made one value, collector current will be beta times the base current. If *RL* is changed, the collector current will *remain the same.* Since we know that Ohm's law must be satisfied, there must be a resistance in the collector circuit that changes as the load resistor is changed. This is R_{CB}, of course.

As long as the transistor is in the active region, R_{CB} will adjust itself to a value that will satisfy the foregoing situation. Therefore, there must be an expression that satisfies this relationship. That is, there must be a mathematical relationship to specify the value of R_{CB} for a specific set of circuit constants. To derive a suitable equation, we must first write the current equations for both the base and the collector circuits in such a manner as to include R_{CB} where applicable. These two equations are chosen because, in the fixed-bias circuit, the value of base current is known (or can be easily found) and the collector-circuit equation is the one that must be solved. In addition, these two equations are related to each other by a simple but important factor.

$$I_C = \beta I_B$$

Hence, the base-current equation and the collector-current equation can be stated as follows. Since

$$I_B = \frac{|V_{CC}|}{RB + R_{BE}} \cong \frac{|V_{CC}|}{RB}$$

and

$$I_C = \frac{|V_{CC}|}{RL + R_{CB} + R_{EE}} \cong \frac{|V_{CC}|}{RL + R_{CB}}$$

these two equations can be combined into a single expression. Substituting the expanded form into the equation $I_C = \beta I_B$,

$$\frac{|V_{CC}|}{RL + R_{CB}} = \beta \times \frac{|V_{CC}|}{RB}$$

We now have an equation that consists of all quantities that are normally known except for R_{CB}. It is now a simple matter to solve for R_{CB}.

First, invert both sides of the equation to take R_{CB} out of the denominator.

$$\frac{RL + R_{CB}}{|V_{CC}|} = \frac{1}{\beta} \frac{RB}{|V_{CC}|}$$

Multiply both sides of the equation by V_{CC}.

$$RL + R_{CB} = RB/\beta$$

This equation tells us that *the load resistor, plus the collector resistance R_{CB}, must be less than RB by the factor β*. The significance of these ideas is simply this: if we know all values of the equation except one, we can solve for this unknown by simple algebraic manipulation. Usually, we know *RL, RB*, and beta. Thus we can easily solve for the value of R_{CB} and then continue the analysis for the dc state of the circuit, because, once R_{CB} is known, the rest of the analysis is accomplished by simple Ohm's law.

Let us solve the foregoing equation for R_{CB} by subtracting RL from each side.

$$R_{CB} = \frac{RB}{\beta} - RL$$

This, then, is the ERCA equation for the simple fixed-bias circuit.

Substituting values in the example,

$$R_{CB} = \frac{200 \text{ k}\Omega}{100} - 1 \text{ k}\Omega = 2 \text{ k}\Omega - 1 \text{ k}\Omega = 1 \text{ k}\Omega$$

We have very simply determined the value for R_{CB}, and we note that its value is equal to *RL*. This indicates that the circuit gamma is 0.5, which is excellent biasing for a linear amplifier. Since this is a beta-dependent circuit, we must, of course, know beta in order to perform the calculation. That is true of any analysis system, and this one is no different.

To see how these ideas affect some practical circuit considerations, we shall put two transistors with different values of beta in the same circuit. Figure 9-1a uses a transistor with a beta of 100, and the values given for V_C and R_{CB} show that the transistor is biased at the midpoint of the dc load line. If the transistor has a different beta, however, as shown in Fig. 9-1b, the circuit values are much different. The transistor is now biased far toward cutoff, and the difference is due *only* to the different value of beta, since RB, RL, and V_{CC} are the same.

Note that the value of R_{CB} is quite different in each case. The value of R_{CB} will adjust itself to a value consistent with the ERCA equation for any given circuit. Since the transistor is replaced by one of lower beta value, its R_{CB} is greater than the unit with higher beta. Since V_{CE} depends upon how R_{CB} compares with RL, and V_{CE} defines the Q point, the Q point depends upon the transistor beta. Thus we have the name "beta-dependent circuit." We see that, for a given circuit, the value of R_{CB} is completely dependent upon beta. And if we know beta, we can solve for any value of R_{CB} that the circuit and the transistor create. We do this by solving the basic ERCA equation for R_{CB}. As stated before, simple Ohm's law will carry us through the rest of the analysis.

Let us now work out a complete dc analysis of a typical circuit using the foregoing ideas. The circuit in Fig. 9-1b will serve, if we assume that we do not yet know the dc values. The first step in the analysis is to draw the dc equivalent circuit and to write the basic ERCA equation.

$$RL + R_{CB} = RB/\beta$$

and solving for R_{CB},

$$R_{CB} = RB/\beta - RL$$

Substituting values,

$$R_{CB} = \frac{200 \text{ k}\Omega}{20} - 1 \text{ k}\Omega = 10 \text{ k}\Omega - 1 \text{ k}\Omega = 9 \text{ k}\Omega$$

The dc equivalent circuit can now be completely filled in with a value for each component, as shown. We can proceed to do the rest of the simple dc analysis. First, collector current can be determined.

$$I_C = \frac{-V_{CC}}{RL + R_{CB}} = \frac{-10}{1 \text{ k}\Omega + 9 \text{ k}\Omega} = \frac{-10}{10 \text{ k}\Omega} = -1 \text{ mA}$$

The collector voltage $-V_C$, referred to ground, is V_{CC} less the drop across the load resistor E_{RL}.

$$E_{RL} = I_C \times RL = 0.001 \times 1 \text{ k}\Omega = 1 \text{ V}$$

$$-V_C = -V_{CC} + E_{RL} = -10 + 1 = -9 \text{ V}$$

Base current is simply V_{CC} divided by RB, if the small junction drop is ignored.

$$-I_B \cong \frac{-V_{CC}}{RB} = \frac{-10}{200 \text{ k}\Omega} = -50 \text{ } \mu\text{A}$$

The gamma can be determined by dividing the total resistance in the collector and emitter circuit into R_{CB}. Since there is no R_E, this reduces to $RL + R_{CB}$.

$$\gamma = \frac{R_{CB}}{RL + R_{CB}} = \frac{9 \text{ k}\Omega}{1 \text{ k}\Omega + 9 \text{ k}\Omega} = \frac{9 \text{ k}\Omega}{10 \text{ k}\Omega} = 0.9$$

With a gamma of 0.9, the transistor is biased very nearly at cutoff.

Note that we can use R_{CB} for the determination of the circuit gamma, rather than V_{CE}. Either is valid, and the choice is determined simply by which is known.

Another use for the ERCA equations is to determine the beta of a transistor by making one simple measurement. If we measure V_{CE}, we can calculate R_{CB}. By solving the basic ERCA equation for β, we can easily determine this for a given transistor. The curcuit of Fig. 9-1b can be solved for beta as shown.

$$\beta = \frac{RB}{RL + R_{CB}} = \frac{200 \text{ k}\Omega}{1 \text{ k}\Omega + 9 \text{ k}\Omega} = 20$$

We have noted in this circuit that the beta of the transistor determines the quiescent operating point. For this reason, the circuit is seldom encountered in practice. It is impossible to make the circuit respond the same for a group of transistors, and usually circuit designers avoid this circuit. However, this circuit *is* useful in describing some of the basic characteristics of the subject of transistor circuit analysis.

9-2 CIRCUIT b—FIXED-BIAS CIRCUIT WITH EMITTER RESISTANCE

A somewhat more complex circuit is shown in Fig. 9-2. This is essentially the same circuit as just used, except for the addition of *RE*, which is usually added for temperature stabilization. (For signal conditions it is often bypassed.) Usually, the analysis of this circuit is

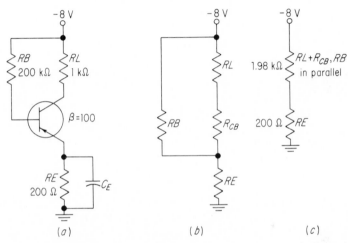

FIG. 9-2. (*a*) Fixed-bias circuit with *RE*; (*b*) and (*c*), dc equivalent circuit.

complicated by the fact that I_B depends to some extent on V_E, and V_E depends upon I_E, which in turn depends upon I_B. Although it is not impossible to determine the dc condition of this circuit by our previous method, it is quite involved. By using the basic ERCA equation for this circuit we shall more easily find the values for this circuit. The ERCA equation for this circuit is derived in the same manner as in the previous example, taking into account the voltage drop across the emitter resistor. This results in the ERCA equation for this circuit which is the same as for the previous one.

$$R_{CB} = \frac{RB}{\beta} - RL = \frac{200 \text{ k}\Omega}{100} - 1 \text{ k}\Omega = 2 \text{ k}\Omega - 1 \text{ k}\Omega = 1 \text{ k}\Omega$$

With the value for R_{CB} determined, we can now proceed to calculate the dc values for the circuit. First, the equivalent circuit shown must be simplified to arrive at the total current through RE. Thus we must find the total parallel resistance of the two parallel branches.

$$R_t = \frac{RB \times (RL + R_{CB})}{RB + (RL + R_{CB})} = \frac{200 \text{ k}\Omega \times 2 \text{ k}\Omega}{200 \text{ k}\Omega + 2 \text{ k}\Omega} = 1.98 \text{ k}\Omega$$

Because RB is so very much larger than the sum of RL and R_{CB}, we could neglect it. However, we shall include it this time to show the more exact method. Now the simplified equivalent circuit is applicable, and can easily be solved for total current.

$$I_t = \frac{-V_{CC}}{R} = \frac{-8}{1.98 \text{ k}\Omega + 0.2 \text{ k}\Omega} = \frac{-8}{2.18 \text{ k}\Omega} = -3.67 \text{ mA}$$

Now that we know the emitter current, we can find the emitter voltage.

$$-V_E = -L_t \times RE = -3.67 \text{ mA} \times 200 = -0.73 \text{ V}$$

Collector current can now be calculated, if we do not wish to consider that $I_C \cong I_E$.

$$-I_C = \frac{-V_{CC} + V_E}{RL + R_{CB}} = \frac{-7.27}{2 \text{ k}\Omega} = -3.63 \text{ mA}$$

The drop across the load resistor will allow us to determine the value of V_C.

$$E_{RL} = I_C RL = 3.63 \text{ mA} \times 1 \text{ k}\Omega = 3.63 \text{ V}$$

$$-V_C = -V_{CC} + E_{RL} = -8 + 3.62 = -4.37 \text{ V}$$

To determine whether the transistor is biased properly, we can calculate for gamma.

$$\gamma = \frac{R_{CB}}{R_t} = \frac{1 \text{ k}\Omega}{2.2 \text{ k}\Omega} = 0.45$$

9-3 CIRCUIT c—SELF-BIAS CIRCUIT

Our third example of the beta-dependent circuits is shown in Fig. 9-3. In this circuit, the base resistor is returned to the collector rather than to V_{CC}. This affords both dc and ac feedback, and the effect is to reduce the signal gain as well as improve the temperature stability. The circuit has the further advantage of reducing distortion, and so it is often used. However, the circuit is very difficult to analyze for its dc operating point. The amount of base current that flows quiescently

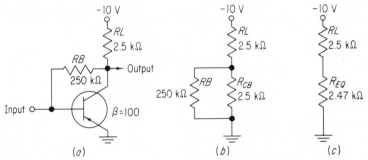

FIG. 9-3.   (a) Self-bias circuit; (b) and (c), dc equivalent circuit.

is a function of V_C, while V_C is a function of collector current, which is dependent upon I_B. Using the methods of Chap. 6, it is not possible to determine the Q point of this circuit, However, using the proper ERCA equation for this circuit, we can very easily calculate for the Q point and all the dc values.

Again, the ERCA equation can easily be derived. It must be realized, however, that base current is determined by the total supply voltage, less the drop across the load resistor; hence the ERCA equation is different from the two preceding examples. The load resistor is not in the beta-dependent "loop," and it must be omitted from the expression. The basic relationship is this: R_{CB} is smaller than RB by the factor beta.

$$R_{CB} = \frac{RB}{\beta} = \frac{250 \text{ k}\Omega}{100} = 2.5 \text{ k}\Omega$$

Now that the proper value for R_{CB} is known, the dc equivalent circuit can be solved. First, the parallel resistance of RB and R_{CB} is determined.

$$R_t = \frac{RB \times R_{CB}}{RB + R_{CB}} = \frac{250 \text{ k}\Omega \times 2.5 \text{ k}\Omega}{250 \text{ k}\Omega + 2.5 \text{ k}\Omega} = \frac{6.25 \times 10^8}{252.5 \times 10^3} = 2.47 \text{ k}\Omega$$

Again note that, with RB much larger than R_{CB}, we should make very little error by simply saying that this equivalent resistance is 2.5 rather than 2.47 kΩ.

Now that a value for R_{CB} is known, we can calculate for all dc conditions of the circuit.

$$-I_{RL} = \frac{-V_{CC}}{RL + R_t} = \frac{-10}{2.5 \text{ k}\Omega + 2.47 \text{ k}\Omega} = -2.01 \text{ mA}$$

$$E_{RL} = I_{RL} \times RL = 2.01 \text{ mA} \times 2.5 \text{ k}\Omega = 5.03 \text{ V}$$

$$-V_C = -V_{CC} + E_{RL} = -10 + 5.03 = -4.97 \text{ V}$$

$$\gamma = \frac{R_{CB}}{RL + R_{CB}} = \frac{2.5 \text{ k}\Omega}{5 \text{ k}\Omega} = 0.5$$

By using the ERCA equations, we have analyzed this circuit for its dc operating point, even though it would normally be considered a difficult thing to do. It turns out to be no more difficult than the very simplest circuit.

9-4 CIRCUIT *d*—SELF-BIAS CIRCUIT WITH EMITTER RESISTANCE

A final example of the beta-dependent circuits is shown in Fig. 9-4. As in the previous example, the analysis of the dc operating point is normally rather difficult, since there are so many interdependent factors. The amount of base current depends upon V_C, which depends upon collector and emitter current, which depend upon base

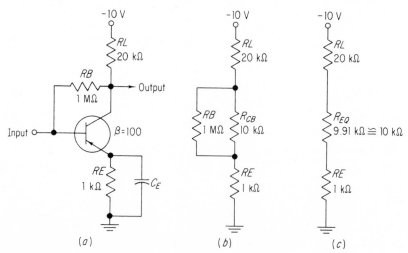

FIG. 9-4. (*a*) Self-bias circuit with *RE;* (*b*) and (*c*), dc equivalent circuit.

current, and so on. With so many variables, each depending upon the other, there is no logical starting point for the analysis. However, by using the ERCA equations we shall have a valid starting point. The derivation of the ERCA equation for this circuit is similar to that of the simple self-bias circuit, except for the fact that the drop across both the emitter resistor and the collector load resistor must be subtracted from the supply voltage to find base- and collector-current values.

The basic relationship for this circuit is that the base resistor RB must be beta times larger than R_{CB}. Solving this relation for R_{CB}, since this is the only unknown in the dc equivalent circuit, gives us

$$R_{CB} = \frac{RB}{\beta} = \frac{1 \text{ M}\Omega}{100} = 10 \text{ k}\Omega$$

In this example we shall neglect the fact that RB shunts R_{CB}, since RB is so very much larger. This will result in an error of about 1 percent, which can easily be tolerated, since the resistors used are generally no better than ±5 percent. The current in the collector and emitter is determined by the total resistance in these circuits and V_{CC}.

$$-I_t = \frac{-V_{CC}}{R_t} = \frac{-10}{31 \text{ k}\Omega} = -0.323 \text{ mA}$$

Now E_{RL} can be found, because we know the current through the load resistor and we know the value of the load resistor itself.

$$E_{RL} = |I_C| \times RL = 0.323 \text{ mA} \times 20 \text{ k}\Omega = 6.45 \text{ V}$$

V_C, the collector voltage measured to ground, can be calculated as shown:

$$-V_C = -V_{CC} + E_{RL} = -10 + 6.45 = -3.55 \text{ V}$$

To find the value of V_{CE}, we must first find the drop across the emitter resistor.

$$-E_{RE} = -I_E \times RE = -0.323 \text{ mA} \times 1 \text{ k}\Omega = -0.323 \text{ V} \quad \text{where } I_E \cong I_C$$

V_{CE} is the drop across R_{CB}, and of course is a function of the collector and emitter current and the drop across the load and emitter resistors.

$$-V_{CE} = -V_{CC} + (E_{RL} + E_{RE}) = -10 + (6.45 + 0.323) = -3.23 \text{ V}$$

To complete the dc analysis of this circuit, we must locate the point along the dc load line that describes the operating condition of the transistor. Gamma will allow us to visualize this point without using the collector curves themselves.

$$\gamma = \frac{V_{CE}}{V_{CC}} = \frac{3.2}{10} = 0.32 \qquad \text{or} \qquad \gamma = \frac{R_{CB}}{R_t} = \frac{10 \text{ k}\Omega}{31 \text{ k}\Omega} = 0.32$$

The transistor is biased at the point that is two-thirds of the way up the load line from cutoff.

BETA-INDEPENDENT CIRCUITS

Any form of analysis for the beta-dependent circuits requires that beta be known, as does the system used in this book. The other class of circuits, the beta-independent circuits, does *not* depend upon beta for the operating point if the circuit is well designed. That is, the value of beta need not be known. It will be noted in the following examples that it is not required to know the value of beta to determine all dc values in the circuit. This is a rather important point, since one would normally expect to use beta in arriving at any transistor-circuit value. Other systems of analysis do require that beta be known, and since beta varies so widely from transistor to transistor, these systems are not accurate unless the beta for each transistor is specifically measured. By using the ERCA equations, however, we can easily — and to a high degree of accuracy — determine the dc condition of any beta-independent circuit.

9-5 CIRCUIT e—THE UNIVERSAL CIRCUIT

In Fig. 9-5, we show a typical beta-independent circuit: the universal circuit in a common-emitter configuration. The dc equivalent circuit is also shown. This circuit can be solved for its dc values by the methods presented in Chap. 6. But it is more enlightening to use the ERCA equations, and more accurate in this case.

The fundamental relation in this circuit is somewhat different than in the previous cases. Because the emitter-base junction is very nearly a short circuit when it is biased in the forward direction, and because the amount of base current is very small compared with the

other currents, this circuit can be treated in the same way as a
balanced-bridge circuit. If the circuit is considered to be a balanced
bridge, a simple proportion will define the relationships within the
circuit. Therefore we can make the following statement: RB must
be larger than the sum of RL and R_{CB} by the same amount that RB'

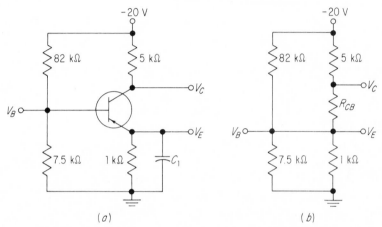

FIG. 9-5.   (a) Universal circuit; (b) dc equivalent circuit.

is greater than RE'. Since we know all values in such a circuit except
R_{CB}, we can set up a simple equation according to this statement and
solve it for R_{CB}.

$$RB : (RL + R_{CB}) :: RB' : RE$$

$$\frac{RB}{RL + R_{CB}} = \frac{RB'}{RE}$$

Cross-multiplying will remove the fractions and allow us to solve
for R_{CB}.

$$RB \times RE = (RL + R_{CB}) \times RB'$$

$$R_{CB} = \left(RB \times \frac{RE}{RB'}\right) - RL$$

This, then, is our basic ERCA equation for this circuit. Using it in the
example of Fig. 9-5, we shall substitute the proper values.

$$R_{CB} = \frac{RB \times RE}{RB'} - RL = \frac{82 \text{ M}\Omega}{7.5 \text{ k}\Omega} - 5 \text{ k}\Omega = 10.9 \text{ k}\Omega - 5 \text{ k}\Omega = 5.9 \text{ k}\Omega$$

Now that a value is known for all resistances in the equivalent circuit, it can be easily solved for all dc values by Ohm's law.

Solving for collector current I_C,

$$-I_C = \frac{-V_{CC}}{RL + R_{CB} + RE} = \frac{-20}{11.9 \text{ k}\Omega} = -1.68 \text{ mA}$$

Once I_C is known, V_C, V_{CE}, and V_E can be determined.

$$-V_C = -I_C(R_{CB} + RE) = -1.68 \text{ mA} \times 6.9 \text{ k}\Omega = -11.6 \text{ V}$$

$$V_{CE} = I_C \times R_{CB} = 1.68 \text{ mA} \times 5.9 \text{ k}\Omega = 9.9 \text{ V}$$

$$-V_E = -I_E \times RE = -1.68 \text{ mA} \times 1 \text{ k}\Omega = -1.68 \text{ V}$$

where $V_B \cong V_E$.

Finally, the gamma of the circuit will allow one to visualize the operating point of this circuit.

$$\gamma = \frac{R_{CB}}{R_t} = \frac{5.9 \text{ k}\Omega}{11.9 \text{ k}\Omega} = 0.496$$

The circuit just described is often used with a transformer in place of the load resistor. If the dc resistance of the primary of the transformer is small, the range of emitter voltage is from ground to $-V_{CC}$ for slow changes. Thus it is possible to severely damage the transistor if the bias is not just right. In this case, the transistor responds very differently to slow changes compared with fast changes. The transistor will not saturate until the base is made to be nearly $-V_{CC}$ for dc conditions, and the collector current will in almost every case be excessive. Therefore special care must be exercised when working with this kind of circuit.

9-6 CIRCUIT f—UNIVERSAL CIRCUIT WITH COLLECTOR FEEDBACK

Circuit f, shown in Fig. 9-6, is sometimes called the *double-feedback circuit* because both RB and RE provide dc and ac feedback. This circuit connection is rather difficult to analyze by conventional means, since there are so many dependent variables. The amount of base current depends on the values of RB and RB′, as well as on the voltage across this divider, V_C. Now V_C depends on collector current, which in turn is a function of base current. Thus there is normally no clear-cut starting point for the analysis. However, all circuit values are known except R_{CB} in the dc equivalent circuit, and the use of the proper

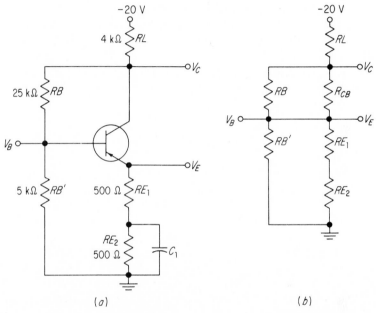

FIG. 9-6. (a) Universal circuit with collector and emitter feedback; (b) dc equivalent circuit.

ERCA equation for this configuration will allow us to determine R_{CB}.

In this circuit, R_{CB} is smaller than RB by the same amount that RE_{total} is less than RB′. Putting this in equation form, we get

$$\frac{R_{CB}}{RB} = \frac{RE_t}{RB'}$$

Cross-multiplying will eliminate the fractions.

$$R_{CB} \times RB' = RB \times RE_t$$

Solving for R_{CB} will give us the ERCA equation.

$$R_{CB} = RB \times \frac{RE}{RB'} = 25 \text{ k}\Omega \times \frac{1 \text{ k}\Omega}{5 \text{ k}\Omega} = 5 \text{ k}\Omega$$

As before, once we know the value of R_{CB}, we can easily solve for all dc conditions in the circuit.

With this circuit, the first thing to be determined is the parallel resistance of the shunt path below the load resistor RL. The current flowing in the base-circuit voltage divider will affect V_C, and so must be included in the total current flow through RL. We must find the parallel resistance of RB and R_{CB}, and RB' and $RE_1 + RE_2$. The shunt resistance of RB and R_{CB} we shall call RP_1, and the shunt resistance of RB' and the two emitter resistors we shall call RP_2.

$$RP_1 = \frac{R_{CB} \times RB}{R_{CB} + RB} = \frac{5 \text{ k}\Omega \times 25 \text{ k}\Omega}{5 \text{ k}\Omega + 25 \text{ k}\Omega} = 4.2 \text{ k}\Omega$$

and where $RE_t = RE_1 + RE_2$,

$$RP_2 = \frac{RB' \times RE_t}{RB' + RE_t} = \frac{5 \text{ k}\Omega \times 1 \text{ k}\Omega}{5 \text{ k}\Omega + 1 \text{ k}\Omega} = 0.83 \text{ k}\Omega$$

Now the total current through the load resistor can be found.

$$-I_t = \frac{-V_{CC}}{RL + RP_1 + RP_2} = \frac{-20}{4 \text{ k}\Omega + 4.2 \text{ k}\Omega + 0.83 \text{ k}\Omega}$$

$$= \frac{-20}{9.03 \text{ k}\Omega} = -2.22 \text{ mA}$$

Then the drop across the load resistor can be determined, and from this we can get a value for V_C.

$$E_{RL} = I_t \times RL = 2.22 \text{ mA} \times 4 \text{ k}\Omega = 8.86 \text{ V} \cong 8.9 \text{ V}$$

With a drop of 8.9 V across RL, the collector voltage can be determined.

$$-V_C = -V_{CC} + E_{RL} = -20 + 8.9 = -11.1 \text{ V}$$

The value of collector current is easily found.

$$-I_c = \frac{-V_C}{R_{CB} + RE} = \frac{-11.1}{5 \text{ k}\Omega + 1 \text{ k}\Omega} = -1.85 \text{ mA}$$

The emitter voltage V_E and the collector-to-emitter voltage V_{CE} can be determined.

$$-V_E = -I_E \times RE = -1.85 \text{ mA} \times 1 \text{ k}\Omega = -1.85 \text{ V}$$

$$V_{CE} = -V_{CC} + E_{RL} + E_{RE} = -20 + 1.85 + 8.86 = -9.29 \text{ V}$$

Gamma will allow us to visualize where on the dc load line the transistor is biased.

$$\gamma = \frac{R_{CB}}{RL + R_{CB} + RE} = 0.5$$

Thus the transistor is biased at the midpoint of the dc load line.

9-7   CIRCUIT g – DUAL-SUPPLY UNIVERSAL CIRCUIT

The method of biasing shown in Fig. 9-7 is widely used. It is an excellent method, having good temperature stability and requiring a minimum of parts. This circuit is intimately related to the universal circuit shown earlier. In one sense it *is* the universal circuit when used with two power supplies, with the collector returned to $-V_{CC}$ and the emitter returned to V_{EE}. The base-circuit voltage divider of the universal circuit must provide some base voltage between $-V_{CC}$

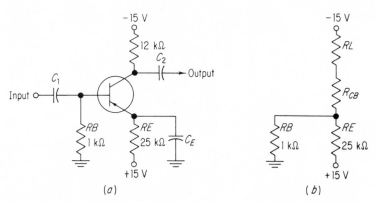

FIG. 9-7.   (*a*) Dual-supply universal circuit; (*b*) dc equivalent circuit.

and V_{EE}. Ground is one of the intermediate voltages between the two supply voltages. Since ground is available without the use of a voltage divider, none is used. The base is simply returned to ground through a single resistor.

The circuit components have dual roles, unlike any others shown thus far. The load resistor RL and the emitter resistor RE are the true biasing resistors, in addition to performing their usual roles. In other words, the biasing is accomplished by the values of RL and RE, while RB has virtually no bearing on this circuit condition. RB serves only as a return to ground for both the base of the transistor and the signal source.

The ERCA equation for the dual-supply circuit is different from any presented thus far. Because the base current is very small relative to collector and emitter current, the emitter of the transistor is essentially at ground potential. Effectively, therefore, there are two separate circuits: I_E and RE, and I_C and $RL + R_{CB}$. There is, however, a commonality that relates each separate circuit to the other. This common factor is the fact that collector and emitter currents are essentially identical. That is, $I_C = \alpha I_E$, but since alpha is very nearly unity,

$$I_C \cong I_E$$

The separate equations are

$$I_C = \frac{|V_{CC}|}{RL + R_{CB}} \quad \text{and} \quad I_E = \frac{|V_{EE}|}{RE}$$

$$\frac{|V_{CC}|}{RL + R_{CB}} \cong \frac{|V_{EE}|}{RE}$$

Solving this f   ression.

$$R_{CB} = \frac{|V_{CC}| \times RE}{|V_{EE}|} - RL$$

In the example, the value of R_{CB} can be found by substituting values.

$$R_{CB} = \frac{RE \times V_{CC}}{V_{EE}} - RL = \frac{25 \text{ k}\Omega \times 15}{15} - 12 \text{ k}\Omega = 13 \text{ k}\Omega$$

Now that R_{CB} has been solved for, it becomes very simple to solve for the dc condition of the circuit. If collector current is to be determined, we can accomplish this as follows:

$$-I_C = \frac{-V_{CC} + V_{EE}}{R_{total}} = \frac{-30}{25 \text{ k}\Omega + 12 \text{ k}\Omega + 13 \text{ k}\Omega} = \frac{-30}{50 \text{ k}\Omega} = -0.6 \text{ mA}$$

where $I_C \cong I_E$

$-V_C$ is easily determined.

$$E_{RL} = I_C \times RL = 0.6 \text{ mA} \times 12 \text{ k}\Omega = 7.2 \text{ V}$$

$$-V_C = -V_{CC} + E_{RL} = -15 + 7.2 = -7.8 \text{ V} \qquad \text{(relative to ground)}$$

The relation of R_{CB} to the total resistance in the *collector circuit* will determine the circuit gamma.

$$\gamma = \frac{13 \text{ k}\Omega}{13 \text{ k}\Omega + 12 \text{ k}\Omega} = 0.52$$

Thus the transistor is biased very nearly at the center of the dc load line. Note that we do not use V_{EE} or RE in this calculation. The reason is that the base is effectively clamping the emitter to ground if RB is reasonably small.

9-8 SILICON NPN TRANSISTORS

Quite often circuits encountered by the average technician use germanium transistors. In these cases it is valid to make the assumption that V_{BE} is negligible. This is the assumption that we have made so far. However, when one encounters silicon NPN units, this is not necessarily the case. Here the emitter-base voltage is much greater, typically 0.7 V, and for good accuracy it must be included in the analytical procedure. In some instances V_{BE} can exceed 1 V, which is a large percentage of typical power-supply voltages. Our equivalent circuit for the dc condition of the transistor must then be altered to include V_{BE} for silicon transistors.

Figure 9-8 shows an equivalent circuit that includes R_{EE}, which, as far as the dc state of the circuit is concerned, we have considered negligible. Again, this is not a strictly accurate equivalent circuit, but its use will greatly facilitate the analysis of circuits using silicon NPNs.

The value of the voltage from base to emitter, V_{BE}, varies somewhat from unit to unit. A typical value, however, is 0.7 V. If the actual value is between 0.4 and 1 V, the use of 0.7 as an average yields very satisfactory results.

V_{BE} remains fairly constant with changing emitter current,

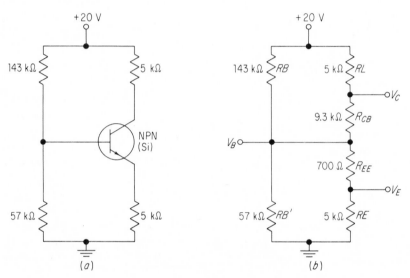

FIG. 9-8.   (a) Silicon NPN example; (b) dc equivalent circuit.

varying only about $\frac{1}{10}$ V at reasonable values of emitter current. Thus we can consider it to be unchanging for values of emitter current in the vicinity of 1 mA or so. Because V_{BE} is nearly constant, R_{EE} varies inversely as the emitter current. It follows, then, that we can use a rule of thumb that will allow us to estimate a value of R_{EE} very closely. R_{EE} is approximately equal to V_{BE} divided by the emitter current at the Q point.

$$R_{EE} \cong \frac{0.7}{I_E \text{ (A)}} = \frac{700}{I_E \text{ (mA)}} \qquad \text{for silicon NPNs only}$$

Thus, if the quiescent value of emitter current is 1 mA, R_{EE} is 700/1, or 700 Ω. The dc equivalent circuit can now be solved by introducing R_{EE} into the ERCA equations. We shall show the dc analysis of a typical universal circuit, leaving the solution of the other six bias methods to the reader. This should pose no problems, since the inclusion of

R_{EE} is the same in every case. R_{EE} and RE are simply summed together and substituted in the basic ERCA equation in place of RE.

Figure 9-8 shows the circuit we shall use to illustrate the method. The basic ERCA relation for this circuit is this: RB is greater than RL plus R_{CB} by the same amount that RB' is greater than R_{EE} plus RE.

$$\frac{RB}{RL + R_{CB}} = \frac{RB'}{R_{EE} + RE}$$

Cross-multiplying and solving for R_{CB} yield

$$RB \times (R_{EE} + RE) = RB' \times (RL + R_{CB})$$

$$R_{CB} = \frac{RB(R_{EE} + RE)}{RB'} - RL$$

Now, unfortunately, we have two unknowns, R_{EE} and R_{CB}. However, since R_{EE} is small compared with the other resistor values, we need only estimate its value. A good idea of its value can be had by first estimating the value of quiescent emitter current. Again using the concept of circuit limits, we can determine the maximum emitter current that would flow *if the transistor were in saturation.*

$$I_{E(\text{max})} = \frac{V_{CC}}{RL + RE} = \frac{20}{10 \text{ k}\Omega} = 2 \text{ mA}$$

If the transistor were biased at a gamma of 0.5, the emitter current would be one-half of 2 mA, or 1 mA. We do not know that this is true at this point, but it is a good guess.

Now an approximate value for R_{EE} can be easily determined.

$$R_{EE} \cong \frac{700}{1} = 700 \ \Omega$$

With an estimated value of resistance for R_{EE}, we can solve for the quiescent values of the circuit.

$$R_{CB} = \frac{RB(R_{EE} + RE)}{RB'} - RL$$

$$= \frac{143 \text{ k}\Omega \ (0.7 \text{ k}\Omega + 5 \text{ k}\Omega)}{57 \text{ k}\Omega} - 5 \text{ k}\Omega$$

$$= 14.3 \text{ k}\Omega - 5 \text{ k}\Omega = 9.3 \text{ k}\Omega$$

At this point, all dc values are known, or can easily be solved by the usual means. The calculated values for this circuit are as follows, and can be verified by the reader:

$$V_B = 5.7 \text{ V} \qquad I_E = 1 \text{ mA}$$
$$V_C = 15 \text{ V} \qquad V_{CE} = 10 \text{ V}$$
$$V_E = 5 \text{ V} \qquad \gamma = 0.47$$

(The remaining determinations for the large-signal and the small-signal response are calculated in the same way as for germanium transistors.)

9-9 GAMMA EQUATIONS

It is oftentimes desirable to be able to determine the circuit gamma (γ) without need for a complete analysis procedure. It is possible to find the value of gamma in one easy step by suitably modifying the ERCA equations. Each of the seven basic biasing circuits can be treated in this manner, as shown in the following discussion.

The first example is the fixed-bias circuit illustrated in Fig. 9-1b; the circuit gamma can quickly be determined as follows.

$$\gamma = \frac{RB/\beta - R_L}{RB/\beta} = \frac{200 \text{ k}\Omega/20 - 1000}{200 \text{ k}\Omega/20} = \frac{9 \text{ k}\Omega}{10 \text{ k}\Omega} = 0.9$$

The circuit is therefore known to be biased very close to cutoff, and this information is derived in one easy step.

The derivation of this relationship is interesting and informative. First, the basic gamma relationship is as follows.

$$\gamma = \frac{V_{CE}}{V_{CC}} = \frac{R_{CB}}{RL + R_{CB} + RE} = \frac{R_{CB}}{R_{\text{total}}}$$

The determination of V_{CE} involves several steps; hence this is not a suitable relationship. But it is possible to modify the second equation since RL and RE are always known, and R_{CB} can be expressed in several different ways. (RE does not exist in this circuit, of course.) The basic ERCA equation for the fixed-bias circuit is

$$R_L + R_{CB} = RB/\beta$$

This can be rewritten.

$$R_{CB} = RB/\beta - RL$$

Properly combining these, and substituting in the basic gamma equation, will provide the required expression.

$$\gamma = \frac{R_{CB}}{RL + R_{CB}} = \frac{RB/\beta - RL}{RB/\beta}$$

This, then, is the gamma equation for the fixed-bias circuit.

If an emitter resistor exists, as in Fig. 9-2, the circuit gamma can as easily be found.

$$\gamma = \frac{RB/\beta - RL}{RB/\beta + RE} = \frac{(200 \text{ k}\Omega/100) - 1 \text{ k}\Omega}{200 \text{ k}\Omega/100 + 200}$$
$$= \frac{2 \text{ k}\Omega - 1 \text{ k}\Omega}{2 \text{ k}\Omega + 0.2 \text{ k}\Omega} = 0.4545 \ldots$$

In the case of the self-bias circuit, the equation is simply

$$\gamma = \frac{RB/\beta}{RL + RB/\beta}$$

Using Fig. 9-3 as an example,

$$\gamma = \frac{RB/\beta}{RL + RB/\beta} = \frac{250 \text{ k}\Omega/100}{250 \text{ k}\Omega/100 + 2.5 \text{ k}\Omega} = 0.5$$

Figure 9-4 illustrates a self-bias circuit with an emitter resistor, and this too can easily be analyzed for gamma

$$\gamma = \frac{RB/\beta}{RL + RB/\beta + RE} = \frac{1 \text{ k}\Omega/100}{20 \text{ k}\Omega + 1 \text{ k}\Omega/100 + 1 \text{ k}\Omega} = 0.32258$$

The gamma equation for the universal circuit is derived by similar reasoning. Since $\gamma = R_{CB}/(RL + R_{CB} + RE)$ and $R_{CB} = (RB \times RE/RB') - RL$, the following holds.

$$\gamma = \frac{(RB \times RE/RB') - RL}{RL + RE + (RB \times RE/RB') - RL}$$

Using the circuit of Fig. 9-5 as an application example, the circuit gamma is easily found.

$$\gamma = \frac{(82 \text{ k}\Omega \times 1 \text{ k}\Omega/7.5 \text{ k}\Omega) - 5 \text{ k}\Omega}{5 \text{ k}\Omega + 1 \text{ k}\Omega + (82 \text{ k}\Omega \times 1 \text{ k}\Omega/7.5 \text{ k}\Omega) - 5 \text{ k}\Omega}$$

$$= \frac{(82 \text{ M}\Omega/7.5 \text{ k}\Omega) - 5 \text{ k}\Omega}{6 \text{ k}\Omega + (82 \text{ M}\Omega/7.5 \text{ k}\Omega) - 5 \text{ k}\Omega}$$

$$= 0.497$$

This is verified by the original analysis accompanying the figure.

The gamma equation for the universal circuit with collector feedback can also be as easily derived. Because for this circuit, $R_{CB} = RB \times RE/RB'$ and $\gamma = R_{CB}/(RL + R_{CB} + RE)$, we can write

$$\gamma = \frac{R_{CB}}{RL + R_{CB} + RE} = \frac{RB \times RE/RB'}{RL + (RB \times RE/RB') + RE}$$

Figure 9-6 will serve as an example.

$$\gamma = \frac{25 \text{ k}\Omega \times 1 \text{ k}\Omega/5 \text{ k}\Omega}{4 \text{ k}\Omega + (25 \text{ k}\Omega + 1 \text{ k}\Omega)/5 \text{ k}\Omega + 1 \text{ k}\Omega}$$

$$= \frac{25 \text{ M}\Omega/5 \text{ k}\Omega}{5 \text{ k}\Omega + 25 \text{ M}\Omega/5 \text{ k}\Omega}$$

$$= \frac{5 \text{ k}\Omega}{5 \text{ k}\Omega + 5 \text{ k}\Omega} = \frac{5 \text{ k}\Omega}{10 \text{ k}\Omega} = 0.5$$

This again agrees with the more conventional analysis accompanying Fig. 9-6.

Finally, the dual-supply configuration can be treated in a like manner.

$$\gamma = \frac{R_{CB}}{RL + R_{CB}}$$

and

$$R_{CB} = \frac{RE \times |V_{CC}|}{|V_{EE}|} - RL$$

Then,

$$\gamma = \frac{(RE \times |V_{CC}|/|V_{EE}|) - RL}{RL + (RE \times |V_{CC}|/|V_{EE}|) - RL}$$

Using Fig. 9-7 as an example,

$$\gamma = \frac{(25 \text{ k}\Omega \times 15/15) - 12 \text{ k}\Omega}{12 \text{ k}\Omega + (25 \text{ k}\Omega \times 15/15) - 12 \text{ k}\Omega}$$

$$= \frac{25 \text{ k}\Omega - 12 \text{ k}\Omega}{12 \text{ k}\Omega + 25 \text{ k}\Omega - 12 \Omega}$$

$$= \frac{13 \text{ k}\Omega}{12 \text{ k}\Omega + 13 \text{ k}\Omega} = \frac{13 \text{ k}\Omega}{25 \text{ k}\Omega} = 0.52$$

QUESTIONS AND PROBLEMS

9-1 A transistor amplifier is shown. Determine the gamma (V_{CE}/V_{CC}) of the circuit. Show, on the load line given, where the Q point is.

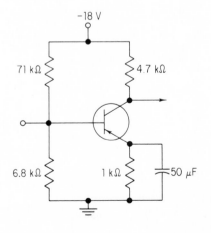
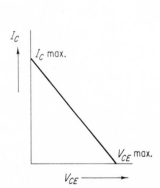

9-2 Given the circuit shown, determine the gamma (V_{CE}/V_{CC}) of the circuit. Show, on the load line given, where Q point is.

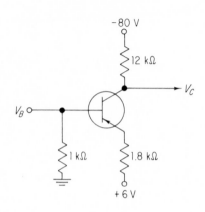
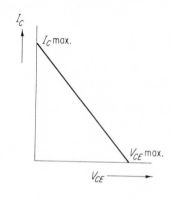

9-3 The transistor shown is a silicon NPN. With the circuit values given, what is the approximate value of V_B?
(a) 3.0 V   (b) 0.07 V
(c) 0.75 V   (d) 0.20 V

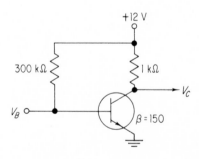

9-4 Draw a dc equivalent circuit of the transistor amplifier shown. Compute the values of R_{CB} and RB if the transistor is biased at the exact center of the dc load line.

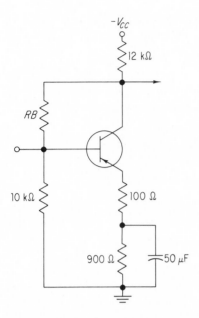

9-5 In the circuit shown, what is the dc equivalent resistance of the collector-base junction?

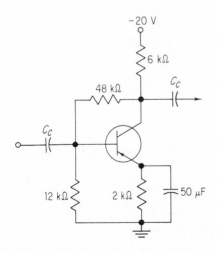

9-6 Discuss briefly how the dc load line sets "limits" upon the transistor.

9-7 In the circuit shown, what is the proper value of RL to bias the transistor at the exact center of the dc load line?

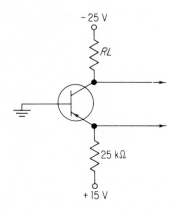

9-8 Refer to Fig. 9-3. Change the value of the collector load resistor to 2.0 kΩ. Calculate the values of R_{CB}; V_C; I_C; I_B.

9-9 Refer to Fig. 9-4. Change the value of RE to 750 Ω. Calculate the values of R_{CB}; V_C; I_C; I_B.

9-10 Refer to Fig. 9-7. Change the value of RE to 18 kΩ. Calculate the values of R_{CB}; gamma; I_C; V_C.

9-11 Refer to Fig. 9-1, and change RL to 1.2 kΩ. Determine R_{CB}; V_{CE}; γ; I_C; I_B.

9-12 Refer to Fig. 9-2, and change RE to 500 Ω. Determine R_{CB}; I_C; I_B; V_{CE}; γ; V_B.

9-13 Refer to Fig. 9-3, and change beta to 75. Find the values for R_{CB}; I_C; I_B; V_{CE}; γ.

9-14   Refer to Fig. 9-4, and change beta to 150. Determine the values of R_{CB}; I_C; I_B; V_{CE}; γ.

9-15   Refer to Fig. 9-5, and change the values of RB and RB' to 51 kΩ and 4.7 kΩ respectively. Find R_{CB}; I_C; I_B; V_{CE}; γ.

9-16   Refer to Fig. 9-6, and change RB to 27 kΩ and RB' to 5.4 kΩ (beta is 100). Find the values of R_{CB}; I_C; I_B; V_{CE}; γ; V_B.

CHAPTER 10 GRAPHICAL ANALYSIS

There are many ways to analyze transistor circuits. For instance, one method may use exact equivalent circuits, with hybrid, T, or π parameters specifying the transistor characteristics. Another method might use a greatly simplified equivalent circuit of the transistor, involving nothing more complex than Ohm's law to determine the circuit operation (the ERCA equations). Still a third method uses the V_{CE}–I_C characteristic curves to help describe the operation of the transistor in a given circuit. Since the graph of the transistor characteristics is used, this method is called *graphical analysis.*

There is no particular preference for one method over another. Whichever one is more applicable in a certain instance is the correct one to use. However, being able to perform a valid graphical analysis of typical circuits will allow one to see more deeply into circuit action, and thus become much more familiar with them.

The methods of graphical analysis are intermediate between the more difficult methods and the very easy methods. The mathematics involved is simple, and one advantage of graphical analysis lies in the fact that the graph itself performs all but the simplest mathematical operations.

The process of graphical analysis, used to describe the functions of a transistor circuit, consists of plotting the dc and signal load lines to determine the "limits" of circuit operation. It is most useful for analyzing large-signal amplifiers, although the process will also provide information relative to small-signal circuits. But, as mentioned, the greatest advantage of performing the methods of graphical analysis is that the student gains a keener insight into circuit operation. One is thus able to work on the circuits in a much more efficient manner. Also, graphical analysis allows one to visualize certain circuit functions, and thus to understand them better.

Because many of the basic circuits have been covered earlier, the general descriptions will be brief. It is assumed that the reader has some familiarity with the basic circuit configurations.

10-1   THE FIXED-BIAS CIRCUIT

Both the circuit and characteristic curves for the first project are shown in Figs. 10-1 and 10-2. The circuit is a simple common-emitter inverter amplifier. Its signal load is labeled r_i, which represents the total input resistance of the next stage of amplification.

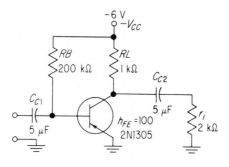

FIG. 10-1.   Fixed-bias circuit for graphical-analysis example.

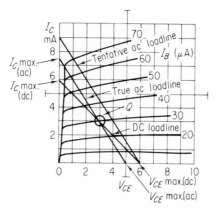

FIG. 10-2.   Load lines for fixed-bias circuit.

    The analysis of this circuit is as simple as the circuit itself, but it will serve to clarify the meanings attached to certain circuit functions. The usual procedure is to describe, first, the dc values that the circuit is producing: I_B, I_C, V_{CE}, etc. Then the large-signal characteristics are determined: $\Delta V_{CE(max)}$, $\Delta I_{C(max)}$, etc. Finally, some useful and practical aspects of small-signal response are approximated, such as voltage gain and input resistance. Once these facts are known, we can predict with good accuracy how the circuit will respond to nearly any signal. Then, if we must troubleshoot the circuit, we can much more easily find the malfunction.

First, we determine the values of the quiescent voltages and currents that exist, as we have done before. These are specified by the two dc limits of the circuit, $V_{CE(max)}$ and $I_{C(max)}$, and by the constants in the circuit itself, RL, RB, etc.

DC Analysis

In this circuit the maximum potential difference between collector and emitter can never exceed the applied voltage, 6 V. This is one extreme limit of the dc load line, and it will be attained ($V_{CE} = V_{CC}$) when the transistor is cut off ($I_C \cong 0$).

The other extreme of the load line occurs when the transistor is fully on, or saturated, and the collector current is limited only by V_{CC} and any series resistance. In this circuit, $I_{C(max)}$ is

$$I_{C(max)} = \frac{V_{CC}}{RL} = \frac{6}{1 \text{ k}\Omega} = 6 \text{ mA}$$

A straight line, drawn between $I_C = 6$ mA on the ordinate and $V_{CE} = 6$ V on the abscissa, is the dc load line. This is shown constructed on the characteristic curves.

Now the quiescent operating point must be found. This can be done in a number of ways, but probably the simplest is to calculate the base current.

$$I_B \cong \frac{V_{CC}}{RB} = \frac{6}{200 \text{ k}\Omega} = 30 \text{ } \mu\text{A}$$

The intersection of the dc load line and the base-current line representing 30 μA is the quiescent operating point, labeled Q on the drawing. From this can be determined the static value of V_{CE} (3 V) and I_C (3 mA). If necessary, I_E can also be derived: $I_E = I_C + I_B = 3$ mA $+ 30$ μA $= 3.03$ mA.

For a very slow change, which amounts to dc, the collector-voltage limits will range between $-V_{CC}$ and ground, a 6-V change. That is, if the base current is slowly increased, the collector voltage will move the Q point toward ground, until, when the transistor is saturated, it falls essentially to ground. As the base current is decreased, the collector voltage rises toward $-V_{CC}$ (-6 V) until, when it reaches this value, it can go no further. Thus, in this simple circuit, the collector voltage can range between $-V_{CC}$ and ground for very slow changes.

We shall shortly find, however, that for fast changes, well within the bandpass of the amplifier, the collector voltage will have a more restricted range, a maximum swing of something less than 6 V. How much less is a function of the slope of the ac, or signal, load line.

Large-Signal Analysis

For this circuit, the signal load line is greatly dependent upon the value of the external load for the transistor. We have previously seen an ac load line for a similar circuit. This one, Fig. 10-2, is constructed in a very simple manner. A *tentative* ac load line is first drawn, from which can be derived the *true* ac load line.

The two limits of the tentative load line are determined by $-V_{CC}$ and the parallel resistance of RL and r_i.

The supply voltage is, of course, -6 V. The total ac load upon the transistor is

$$\text{Load}_{ac} = \frac{RL \times r_i}{RL + r_i} = \frac{1 \text{ k}\Omega \times 2 \text{ k}\Omega}{2 \text{ k}\Omega + 2 \text{ k}\Omega} = 667 \ \Omega$$

The maximum collector current that would flow if the collector voltage could rise to $-V_{CC}$ is

$$I_{C(\max)}(\text{tentative}) = \frac{V_{CC}}{\text{load}_{ac}} = \frac{6}{667} = 9 \text{ mA}$$

A line drawn from $-V_{CC}$ (6 V) to 9 mA is the tentative load line. However, note that the tentative load line does not go through the quiescent operating point. By drawing a third line (the true ac load line) *parallel* to the tentative load line, but passing through the Q point, we can finally specify the actual signal conditions for this circuit. The true maximum collector current for this circuit is about 7.5 mA, while V_{CE} maximum is 5 V.

This simply means that if the transistor is momentarily cut off, the collector voltage will rise to -5 V with respect to ground, but can go no further. With the transistor driven to saturation, V_{CE} can fall to essentially 0 V. Hence, for signal conditions, the collector voltage can range from ground to -5 V, a 5-V total change. For dc conditions (or for very slow changes), the collector-voltage range is 6 V. This is typical of all amplifiers to a greater or lesser degree. The collector-voltage excursion is almost always smaller for signal conditions than it is for dc conditions. The only time the dc and ac load lines coincide is when there is no significant external loading (r_i very high), which in the practical circuit is seldom true.

The large-signal, or maximum-signal, conditions are determined from the ac load line. The maximum collector-voltage range is from essentially ground to −5 V. The maximum collector-current range is

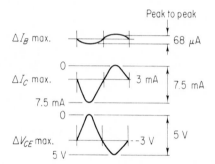

FIG. 10-3.   Nonlinear waveforms.

from essentially 0 to about 7.5 mA. The base-current drive necessary to cause these changes is from 0 to 68 μA. Typical waveforms for these values are shown in Fig. 10-3.

A more meaningful way of showing the waveforms is illustrated in Figs. 10-4 and 10-5. The first shows the result of a moderately

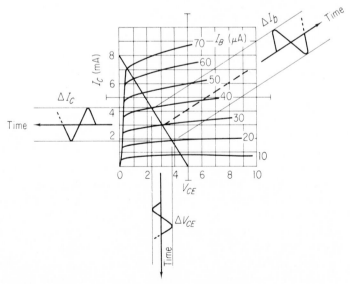

FIG. 10-4.   Indicating current and voltage excursions along the ac load line.

small-signal current impressed upon the circuit. The resultant current (I_C) and voltage (V_{CC}) changes are shown plotted against the signal load line. If this is done carefully, the distortion produced by the transistor can be easily seen. In this case, the excursion is small enough, and the base-current curves are evenly enough spaced, so that little distortion is evident.

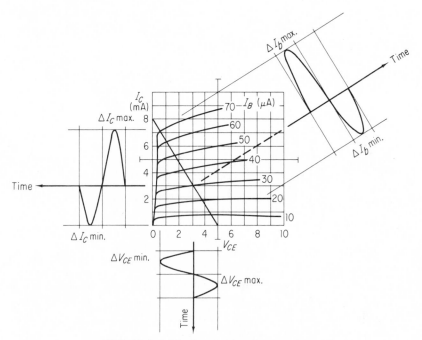

FIG. 10-5.   Maximum signal excursions.

By assuming a larger input (Fig. 10-5), sufficient to drive the transistor just to its limits of operation, the large-signal characteristics of the circuit can be visualized. Clearly evident is the fact that the circuit is not biased at the center of the ac load line. If the circuit were to be redesigned, an exactly symmetrical output could be attained for maximum signal conditions.

Note that if the input amplitude were to be increased still further, the output would be "clipped." In such a case the amplifier would be said to be "overdriven." In a linear circuit this is to be avoided as the most severe form of distortion. Since the power supply now limits the maximum output, the only way to eliminate this for a given-size input is to increase V_{CC} to allow a larger swing either side of the operating point. Thus the circuit designer must ensure that the power supply is adequate for the largest signal to be encountered if clipping distortion is to be avoided.

Small-Signal Analysis

The practical small-signal characteristics of the circuit can now be approximated. Those of concern to us will be the input resistance and voltage gain. Although these are not, strictly speaking, a part of the graphical analysis, they are included for the sake of completeness.

The input resistance can be approximated closely enough for our purposes by considering the ac resistance of the emitter itself, r_e, to be, in this case, about 9 Ω. This is derived from the following relationship:

$$r_e = \frac{26}{I_E \text{ (mA)}} = \frac{26}{3} \cong 9$$

where 26 is a constant derived from the transistor physics.

The input resistance is $(\beta + 1)$ times this value.

$$r_i = (\beta + 1) \, r_e = 101 \times 9 = 909 \, \Omega$$

Because this value is very much smaller than RB, it is for all practical purposes the input resistance. If RB were about the same value, the total input resistance would be the parallel resistance of RB and 909 Ω. In the case of the circuit values shown, the total parallel value would be 904 Ω, which is not enough difference to be concerned about.

The small-signal voltage gain is easily determined by the following relationship.

$$A_v = \frac{(RL \times r_i)/(RL + r_i)}{r_e} = \frac{(1000 \times 2000)/(1000 + 2000)}{9} = \frac{667}{9} = 74$$

Hence a 1-mV input would result in a 74-mV output.

In the case of this simple circuit, the complete analysis for our purposes is therefore quite simple. A more complex example follows, which, while not greatly different from the preceding one, does nevertheless require some further thought.

10-2 THE UNIVERSAL CIRCUIT

The circuit to be used in the present discussion is given in Fig. 10-6, along with collector curves in Fig. 10-7. We shall graphically analyze this circuit for its quiescent values and its large- and small-signal characteristics.

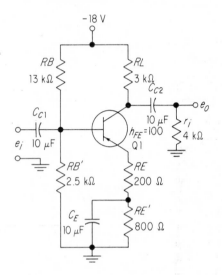

FIG. 10-6. Universal-circuit example.

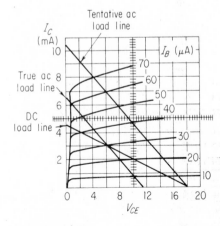

FIG. 10-7. Load lines for universal circuits.

DC Analysis

Our first step is to locate the limits of the dc load line. One limit, of course, is $-V_{CC}$, or 18 V. The other limit is the maximum current that can flow in the collector circuit with the transistor full on.

$$I_{C(max)} = \frac{V_{CC}}{RL + RE' + RE} = \frac{18}{4 \text{ k}\Omega} = 4.5 \text{ mA}$$

The dc load line is shown in Fig. 10-7 drawn between these two limits. The quiescent operating point must now be found, so that it can be placed upon the load line. This is conveniently done as follows:

$$V_B = \left(\frac{RB'}{RB + RB'}\right)(-V_{CC}) = \left(\frac{2.5 \text{ k}\Omega}{2.5 \text{ k}\Omega + 13 \text{ k}\Omega}\right)(-18) = -2.9 \text{ V}$$

Since $V_B \cong V_E$, we can now determine the emitter current. (For dc conditions consider that $RE = 800 + 200 = 1 \text{ k}\Omega$.)

$$I_E = \frac{V_E}{RE} = \frac{2.9}{1000} = 2.9 \text{ mA}$$

Also, $I_C \cong I_E$, and the drop across RL can now be found, which will allow V_{CE} to be calculated.

$$E_{RL} = I_C \times RL = 2.9 \text{ mA} \times 3 \text{ k}\Omega = 8.7 \text{ V}$$

$$V_{CE} = -V_{CC} + E_{RL} + E_{RE} = -18 + 8.7 + 2.9 = -18 + 11.6 = -6.4 \text{ V}$$

The intersection of the dc load line with a V_{CE} of -6.4 V is the Q point, as shown. Hence the collector current must be about 2.9 mA, while base current is about 28 μA. We now have enough information to determine the rest of the static voltages and currents around the circuit. The known values are

$$
\begin{array}{ll}
V_B = -2.9 \text{ V} & V_{CE} = 6.4 \text{ V} \\
V_E \cong -2.9 \text{ V} & E_{RL} = 8.7 \text{ V} \\
I_C \cong 2.9 \text{ mA} & E_{RE} = 0.58 \text{ V} \\
I_E \cong 2.9 \text{ mA}
\end{array}
$$

The collector voltage, referred to ground, and the drop across RE' must still be determined. These are important because we should normally measure the circuit voltages from ground; we should seldom have occasion to measure, for instance, V_{CE}.

Equating the collector loop for V_C yields

$$-V_C = -V_{CC} + E_{RL} = -18 + 8.7 = -9.3 \text{ V}$$

The voltage across the 800-Ω emitter resistor V_E is simply the product of emitter current and the value of the resistor.

$$-V_E = I_E \times RE' = 2.9 \text{ mA} \times 800 = -2.32 \text{ V}$$

Large-Signal Analysis

Knowing all the dc values in the circuit will allow us to begin the construction of the signal, or ac, load line. Unfortunately this is not as simple as in the case of the earlier circuit. For instance, the two limits of operation, saturation and cutoff, will not necessarily result in expected circuit values. When the transistor is momentarily cut off by a signal within its bandpass, the collector-to-emitter voltage will not be V_{CC}. Also, when the transistor is driven to saturation ($V_{CC} \cong 0$ V), the maximum collector current will be dependent upon several factors.

To determine most easily the limits of the signal load line, we shall first have to determine the tentative limits. One of these limits (Q_1 off) is, of course, $-V_{CC}$.

The other limit, $I_{C(max)}$, occurs when the transistor is driven to saturation. Its value is a function of the total applied voltage and the total resistance through which the current must flow. The total signal load as viewed from the collector is the parallel resistance of RL and r_i.

$$\text{Load}_{ac} = \frac{RL \times r_i}{RL + r_i} = \frac{3 \text{ k}\Omega \times 4 \text{ k}\Omega}{3 \text{ k}\Omega + 4 \text{ k}\Omega} = \frac{12 \text{ M}\Omega}{7 \text{ k}\Omega} = 1.71 \text{ k}\Omega$$

$$I_{C(max)} = \frac{V_{CE(max)}}{\text{load}_{ac}} = \frac{18}{1.71 \text{ k}\Omega} \cong 10.5 \text{ mA}$$

This, then, is the other limit of the tentative signal load line. A line drawn between these two limits is the tentative load line. Another line, parallel to this, but passing through the Q point, is the true ac load line.

From this can be derived the actual $V_{CE(max)}$ and $I_{C(max)}$ values. $V_{CE(max)}$ is seen to be about 11.7 V, while $I_{C(max)}$ is on the order of 6.5 mA. As always, a graphical determination cannot provide exact results, but it is surprising how accurate it can be if carefully done.

The ac load line can be arrived at by another method that uses simplified equivalent circuits to obtain the signal limits. We shall need equivalent circuits for each of the following three conditions:

1.   Quiescent operation point
2.   Transistor cut off
3.   Transistor saturated

Although this method is more detailed than the one just accomplished, it gives a keener insight into the "why" of circuit operation.

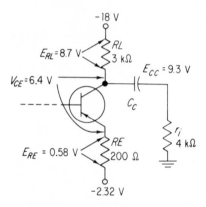

FIG. 10-8.   Voltage drops in collector-emitter circuits.

Figure 10-8 shows the equivalent collector-emitter circuit for
the static condition. Note that the lower end of RE is shown as returning to a source of voltage. In effect, the 800-Ω resistor and the
10-μF capacitor form a separate power supply, for all practical
purposes. (This is of little importance as far as the dc conditions are
concerned, but we shall soon make good use of this idea.) The important voltage drops are shown, all of which total 18 V.

Keeping in mind the fact that the following discussion concerns
the signal conditions, i.e., that the changes are occurring very fast,
let us proceed to the second condition, where the transistor is momentarily cut off. Just as the transistor is driven to cutoff by a fast
signal, the circuit conditions shown in Fig. 10-9 prevail. With the
transistor cut off, there is no current through the transistor; so there
is no drop across RE. However, the large bypass capacitor CE holds
the lower end of RE to −2.32 V; thus the emitter voltage referred to
ground is −2.32 V.

Because the coupling capacitor has a charge upon it due to the
quiescent voltages, when the transistor is quickly driven to cutoff,
current continues to flow as shown. No smaller current than this can
flow in RL, unless the transistor is held off for a longer time. If this

were done, the capacitor would have time to recharge to this new value. But we are concerned only with signal conditions; so we must calculate the value of this current, I_{min}, to determine the minimum possible voltage drop across RL.

Initially, the transistor was operating at its quiescent values, and now the signal has caused it to be driven just to cutoff. The voltage to which the capacitor will try to recharge is that value that would be

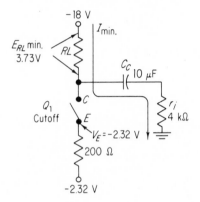

FIG. 10-9.   Minimum current flow in RL for signal conditions.

attained if the transistor were held off for a long time. This voltage is determined as shown below, labeled V_{eff}.

$$V_{eff} = -V_{CC} + E_{CC2} = -18 + 9.3 = 8.7 \text{ V}$$

This is the amount of change that would occur across C_{C2} if the transistor were held off for a long time, and the capacitor would have to recharge to a new value that was 8.7 V different than at the Q point. Hence the least current through RL and r_i is I_{min} (for a fast change).

$$I_{min} = \frac{V_{eff}}{RL + r_i} = \frac{8.7}{7000} = 1.243 \text{ mA}$$

This current, flowing through RL, produces a voltage drop that prevents the collector of the transistor from going completely to $-V_{CC}$ if it is momentarily cut off by the signal.

$$E_{RL(min)} = I_{min} \times RL = 1.243 \text{ mA} \times 3000 = 3.73 \text{ V}$$

Thus, relative to ground, the collector can go no more negative than $-18 + 3.73 = -14.27$ V. However, we must determine $V_{CE(max)}$, to be able to locate one limit of the ac load line.

Equating the collector-emitter loop for V_{CE},

$$V_{CE(max)} = -V_{CC} + E_{RL(min)} + E_{RE'} = -18 + 3.73 + 2.32 = -11.95 \text{ V}$$

This is the maximum value that V_{CE} can rise to, and so it is one limit of the ac, or signal, load line. On the curves of Fig. 10-7 this is clearly indicated.

Now, a straight line drawn through the Q point from $V_{CE} = 11.95$ V is the ac load line. We have not proved that this is a true signal

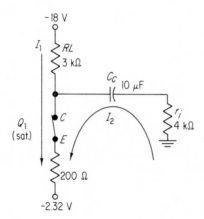

FIG. 10-10. Maximum current flow in the collector for signal conditions.

load line, however. By calculating for the upper extreme, $I_{C(max)}$ (for signal conditions), we can verify that this load line is truly a good one for this circuit.

The second limit of circuit operation is $I_{C(max)}$, and the equivalent circuit shown in Fig. 10-10 can be used to determine its value. Now consider that the transistor has been driven from its Q point to saturation by the signal, a very fast transition.

As shown in the figure, two currents will flow, and each will flow through the transistor. The value of each can be approximated in the following procedure. Following this, a more precise determination is given.

First, let us determine the value of I_1, for this is done in a very straightforward way.

$$-I_1 = \frac{-V_{CC} + E_{RE'}}{RL + RE} = \frac{-18 + 2.32}{3 \text{ k}\Omega + 0.2 \text{ k}\Omega} = \frac{-15.68}{3.2 \text{ k}\Omega} = -4.9 \text{ mA}$$

(The minus sign simply indicates that the current is flowing in a direction caused by the negative voltage.)

The value of I_2 is not so easily determined. However, the same principles used to determine $V_{CE(max)}$ can be used here.

I_2 is the current flowing as the result of V_{CE} changing from Q to saturation, and hence requires that the coupling capacitor recharge toward this new value. Of course, for very rapid changes it will never attain this new value, but it must make the effort.

The voltage to which it will try to charge is equal to the change in V_{CE} that occurs as the transistor goes from its Q point to saturation. At the Q point, $V_{CE} = 6.4$ V, while at saturation $V_{CE} \cong 0$ V. The collector-to-emitter voltage, then, changes by 6.4 V. This is considered a source, and is again labeled V_{eff}

$$I_2 = \frac{V_{eff}}{r_i + RE} = \frac{6.4}{4.2 \text{ k}\Omega} = 1.52 \text{ mA}$$

The total collector current $I_{C(total)}$ is approximately the sum of these.

$$I_{C(total)} = I_1 + I_2 = 4.9 \text{ mA} + 1.52 \text{ mA} = 6.43 \text{ mA}$$

Of course, two currents cannot simply be summed in a circuit such as this, since each will influence the voltages in the circuit and hence change the conditions that determine the other. Probably the simplest method of determining the exact value of current in this circuit is by the use of superposition. The superposition theorem states that a complex circuit can be solved by making a break, or open circuit, at a convenient place in the circuit, thus removing one current path. Then, the effective voltage across this break V_{eff} can be determined, usually by Ohm's law. Now, the true value of current at this point is of a value that this voltage would produce across the effective impedance R_{eff} seen *looking back into the break* with all power sources replaced with their internal impedances (usually 0 Ω).

In the present instance, a convenient place to effect an imaginary open in the circuit is just above the 200-Ω resistor. With the break made, no current flows in this branch, and therefore the voltage across the open circuit can easily be found. First, find the current flowing in the branch containing the 3000- and 4000-Ω resistors.

$$-I = \frac{-18 + 9.3}{7000} = -1.2428571 \text{ mA}$$

The voltage V_C at the collector of the transistor in reference to ground can now be determined.

$$E_{RL} = 1.2428571 \times 3000 = 3.7285713 \text{ V}$$

$$-V_C = -18 + 3.7285713 = -14.271428 \text{ V}$$

(Note: the calculations are not rounded off, in order to illustrate the accuracy of the final result.)

The voltage at the other break point (top of the 200-Ω resistor) is simply -2.32 V, since no current is flowing in this branch. The total voltage V_{eff} across the break is the potential difference between these two values.

$$V_{\text{eff}} = -14.271429 + 2.32 = -11.951428 \text{ V}$$

Now, the impedance seen looking back into the open, with all sources replaced with a short circuit, consists of the 3000- and 4000-Ω resistors in parallel and all in series with the 200-Ω resistor.

$$R_{\text{eff}} = \frac{1}{1/3000 + 1/4000 + 200} = 1714.2857 + 200 = 1914.2857 \ \Omega$$

The true current, then, is

$$I_E = \frac{|V_{\text{eff}}|}{R_{\text{eff}}} = \frac{11.951428}{1914.2857} = 6.2432833 \text{ mA}$$

The approximation, then, is seen to be not greatly in error.

Note the upper termination of the ac load line. It terminates at almost exactly this value of current, and therefore the load line is proved accurate and valid by plotting the third point on the line.

Now that the ac load line is plotted on the curves, by either of the preceding methods, we can use it to determine the large-signal response of the circuit.

The first rather obvious result of constructing the ac load line is to use it to determine the maximum peak-to-peak output voltage. The load line, of course, indirectly gives us this information in terms of collector-to-emitter voltage. We need merely change this to a voltage in terms of actual circuit values. In other words, $\Delta V_{CE(max)}$ must be changed so that this basic information is in terms of collector voltage referred to ground, because this is no doubt the way that we should measure it.

The change in collector voltage, $\Delta V_{C(max)}$, referred to ground, that occurs as the transistor is just driven from saturation to cutoff by the signal, is derived from the curves and the ac load line.

First, assume that the transistor is cut off. V_{CE} is maximum, 11.95 V, and to this must be added the dc voltage across the emitter-bypass capacitor, E_{CE}. This has already been determined as -2.32 V. Therefore $V_{C(max)}$ is the sum of these, and the collector voltage, referred to ground, can never go more negative than $(-11.95) + (-2.32) = -14.27$ V.

The most positive that the collector can go, again referred to ground, is that value produced when the transistor is just driven to saturation. Now our previously determined value of $I_{C(max)}$ can be used again. With V_{CE} essentially equal to zero, I_{max} flowing through RE sets the emitter at some maximum value $(I_C \cong I_E)$.

$$E_{RE(max)} = I_{max} \times RE = 0.00624 \times 200 = 1.248 \text{ V} \cong 1.25 \text{ V}$$

Figure 10-11 shows the equivalent circuit for this condition. Starting at ground (0 V), we can now easily determine the voltage at the emitter V_E.

$$V_E = E_{RE'} + E_{RE} = (-2.32) + (-1.25) = -3.57 \text{ V}$$

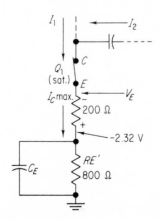

FIG. 10-11. Maximum current flow in the emitter for signal conditions.

Ignoring the drops across the transistor junctions, if V_E is at -3.57 V, the collector V_C is also nearly at a potential of -3.57 V, since the transistor is in saturation ($V_{CE} \cong 0$).

The range of collector voltage, then, for signal conditions, is from -14.27 V maximum to -3.57 V minimum. This is shown graphically in Fig. 10-12, for one complete cycle. Note that this is not a symmetri-

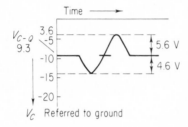

FIG. 10-12.   Illustrating maximum possible collector excursion $\Delta V_{CE(max)}$.

cal waveform. This is due to the fact that the circuit is not biased at the exact center of the ac load line. Since we should seldom drive an amplifier so hard, this small difference is of little consequence. However, if the maximum excursion waveform were *very* unsymmetrical, the circuit would require rebiasing so as to operate nearer the center of the ac load line, to prevent clipping of large signals.

Small-Signal Analysis

An approximate indication of voltage gain and input impedance will allow us to describe the circuit action in greater detail. As before, we shall make a few simplifying assumptions to allow us to easily determine these important characteristics of circuit operation.

In the circuit diagram of Fig. 10-6, the voltage gain is determined by several factors. The dc load, the signal (or ac) load, the emitter-base resistance, and the value of RE, all bear directly upon the voltage gain. The total ac load is the parallel combination of RL and r_i.

$$\text{Load}_{ac} = \frac{RL \times r_i}{RL + r_i} = \frac{3 \text{ k}\Omega \times 4 \text{ k}\Omega}{3 \text{ k}\Omega + 4 \text{ k}\Omega} = 1.714 \text{ k}\Omega \cong 1.71 \text{ k}\Omega$$

The total unbypassed resistance in the emitter lead is the sum of the emitter resistance, r_e, and RE.

$$r_e = \frac{26}{I_E \text{ (mA)}} = \frac{26}{2.9} = 8.9 \cong 9 \text{ }\Omega$$

$$RE_{\text{total}} = RE + r_e = 200 + 9 = 209 \text{ }\Omega$$

With this information at hand, we can determine the voltage gain of the circuit.

$$A_v = \frac{\text{load}_{ac}}{RE_{\text{total}}} = \frac{1.71 \text{ k}\Omega}{209} = 8.2$$

Hence a signal voltage impressed upon the base will be amplified by about 8 times.

Now, the input resistance is, at midband frequencies, the parallel resistance of the two base-biasing resistors, RB_{eq}, and the input resistance of the transistor itself, R_{ib}.

$$RB_{eq} = \frac{RB \times RB'}{RB + RB'} = \frac{13 \text{ k}\Omega \times 2.5 \text{ k}\Omega}{13 \text{ k}\Omega + 2.5 \text{ k}\Omega} = 2.1 \text{ k}\Omega$$

The input resistance of the transistor itself, R_{ib}, is $(\beta + 1)$ times the unbypassed resistance in the emitter lead, which we determined above as 209 Ω. Assume $\beta = 100$.

$$R_{ib} = (\beta + 1)209 = 101 \times 209 = 21.1 \text{ k}\Omega$$

The total input resistance r_i can now be calculated.

$$r_i = \frac{RB_{eq} \times R_{ib}}{RB_{eq} + R_{ib}} = \frac{2.1 \text{ k}\Omega \times 21.1 \text{ k}\Omega}{2.1 \text{ k}\Omega + 21.1 \text{ k}\Omega} = 1.9 \text{ k}\Omega$$

We have accomplished what we set out to do. We have described the dc condition of the circuit, as well as both the large- and small-signal response, and we have done this with a relative minimum of effort. Nearly any simple circuit can be analyzed by the techniques used here. Some circuits, however—notably those with multiple feedback paths—require more advanced and sophisticated methods.

10-3 THE DUAL-SUPPLY CIRCUIT

The circuit shown in Fig. 10-13 is a so-called dual-supply circuit, very similar in many respects to the universal circuit previously described. As shown, the circuit is a common-emitter small-signal voltage amplifier, intended for a bandpass from 5000 Hz to better than 3 MHz. The low-frequency response is deliberately made poor in this instance, as evidenced by the small coupling and bypass capacitors. The high-frequency response is partly determined by the choice of transistor, a 2N711A.

A somewhat more complex version of this circuit is used in several applications; so its thorough study is quite relevant. One of these

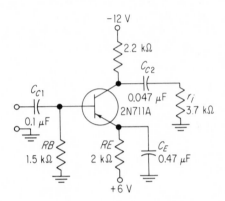

FIG. 10-13. Dual-supply-circuit example.

circuits, taken from actual equipment, will be described in Chap. 12, and the present simplified circuit will form a basis for later discussion.

DC Analysis

As always, we start our investigation of the circuit by finding the dc conditions in the circuit. This particular circuit is slightly different from most. To determine the values for the operating conditions, as well as the signal conditions, will require that we take a slightly different approach.

First, we note that the base connection is returned to ground through a very low-value resistor. Since base current is very small, we shall expect little or no measurable voltage drop across the base resistor RB. It can therefore be assumed that the base voltage, relative to ground, will be essentially zero. By ignoring the small drop across the base-emitter junction, it can also be seen that the emitter voltage is nearly zero with respect to ground.

This observation, then, yields the starting point for determining the quiescent operating point. On the assumption that $V_E = 0$ V, relative to ground, and knowing the value of RE, the emitter biasing resistor, the emitter current flowing through RE can be found.

$$I_E = \frac{E_{RE}}{RE} = \frac{6}{2000} = 3 \text{ mA}$$

In this circuit, if *RB* is not too large, the amount of resistance in the emitter lead is the factor that determines the amount of transistor current. This is true because the emitter voltage is essentially at ground for any reasonable value of *RB;* so simple Ohm's law will dictate emitter current. (If, in the above example, *RE* is changed to a 1-kΩ resistor, emitter current will increase to 6 mA, and the other currents will increase correspondingly.) Hence *RE* is seen to be the true biasing resistor for the circuit.

Once a value is found for emitter current, it is a simple matter to compute base and collector currents. ($\beta = h_{FE} = 50$.)

$$I_C = \alpha I_E = \frac{\beta}{1+\beta} \times I_E = \frac{50}{1+50} \times 3 \text{ mA} = 0.98 \times 3 \text{ mA} = 2.94 \text{ mA}$$

$$I_B = \frac{I_C}{\beta} = \frac{2.94 \text{ mA}}{50} \cong 60 \ \mu A$$

(Note that we could easily say that $I_C \cong I_E = 3$ mA, and the collector current would then be in error by only 60 μA.)

Because the base, and therefore the emitter, can never go far away from ground, we can consider the circuit as though the emitter were tied directly to ground. Hence the dc load line is determined by only $-V_{CC}$ and *RL*.

$$\text{Limit 1:} \quad V_{CE} = V_{CC} = -12 \text{ V}$$

$$\text{Limit 2:} \quad I_{C(\text{max})} = \frac{V_{CC}}{RL} = \frac{12}{2.2 \text{ k}\Omega} = 5.5 \text{ mA}$$

The dc load line is drawn between these two limits, as shown in Fig. 10-14. For purposes of locating the quiescent operating point,

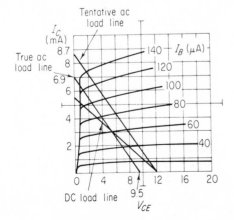

FIG. 10-14. Dual-supply-circuit load lines.

we shall consider that $I_C \cong I_E$, or 3 mA. Thus the intersection of the dc load line and the $I_C = 3$ mA line is the operating point. From this, we can derive V_{CE}, which is about 5.5 V.

The operating point of the circuit and the dc load line, then, are found to be easily determined.

Large-Signal Analysis

Now, to construct the signal, or ac, load line, we must go one step further. With the transistor operating under signal conditions, we can no longer ignore the coupling capacitor C_{c2}, for it appears to be a short circuit to high-frequency signals. Thus the total ac load for the transistor is the shunt resistance of the load resistor, RL, and the input resistance of the following stage, r_i.

$$\text{Load}_{ac} = \frac{RL \times r_i}{RL + r_i} = \frac{2.2 \text{ k}\Omega \times 3.7 \text{ k}\Omega}{2.2 \text{ k}\Omega + 3.7 \text{ k}\Omega} = 1.38 \text{ k}\Omega$$

Now, if the maximum voltage between collector and emitter could rise to 12 V with the transistor just at cutoff, maximum collector current as the transistor is saturated would be simply V_{CC} divided by the ac load.

$$I_{C(\text{max})} = \frac{V_{CC}}{\text{load}_{ac}} = \frac{12}{1.38 \text{ k}\Omega} = 8.7 \text{ mA}$$

A straight line drawn between $-V_{CC}$ and 8.7 mA is the tentative load line. However, $V_{CE(\text{max})}$ cannot rise all the way to V_{CC} when the transistor is cut off; so we must reposition the tentative load line. A line parallel to the tentative load line, but passing through the Q point, is therefore the true ac load line for this circuit. Note that $V_{CE(\text{max})}$ is 9.5 V, while $I_{C(\text{max})}$ is slightly under 7 mA. The large-signal characteristics of this circuit, then, are specified by this signal load line. Maximum collector-to-emitter swing is about 9 V, and collector current can vary by about 7 mA. It requires about 120 μA of base-current drive to swing the transistor from cutoff to saturation.

Small-Signal Analysis

The small-signal analysis for this circuit is relatively simple. We are mainly concerned with only the input resistance (to determine loading on the preceding stage) and the voltage gain.

First, we must determine r_e, the emitter resistance.

$$r_e = \frac{26}{I_{E \, (mA)}} = \frac{26}{3} = 8.7 \; \Omega \cong 9 \; \Omega$$

The input resistance to a signal is $(\beta + 1)$ times this value, approximately, shunted by RB.

$$r_i = \frac{(\beta + 1)r_e \times RB}{(\beta + 1)r_e + RB} = \frac{0.46 \; k\Omega \times 1.5 \; k\Omega}{0.46 \; k\Omega + 1.5 \; k\Omega} = 352 \; \Omega$$

The voltage gain A_v is simply the total ac load divided by the emitter resistance r_e.

$$A_v = \frac{\text{load}_{ac}}{r_e} = \frac{1.38 \; k\Omega}{9} = 153$$

Note that we can ignore RE for signal conditions since it is bypassed.

We have described the dc, large-signal, and small-signal conditions in this circuit. We now know such things as the quiescent voltages and currents, the maximum peak-to-peak collector excursion, and the input resistance and voltage gain. Each of these items can be easily verified with either the volt-ohm-milliammeter for dc values or the oscilloscope for signal conditions.

10-4 THE SELF-BIAS CIRCUIT

When a circuit such as the one shown in Fig. 10-15 is to be analyzed, the graphical methods used up to this point are not adequate. For several reasons, the circuit is rather difficult to work with; so in this chapter we have deliberately avoided this kind of circuit thus far. However, in typical industrial circuits these are frequently used, and one should be able to analyze them graphically to achieve a more thorough understanding of transistor circuitry.

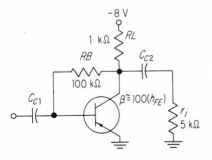

FIG. 10-15. Self-bias-circuit example.

DC Analysis

As in any circuit, the first step in the graphical analysis is to construct the dc load line on the characteristic curves. This is done in Fig. 10-16 in exactly the same manner as with any circuit. The two limits of the dc load line are 8 V ($V_{CE} = V_{CC}$) and 8 mA ($I_{C(max)} = V_{CC}/RL$). The dc load line is shown drawn between these two limits in Fig. 10-16.

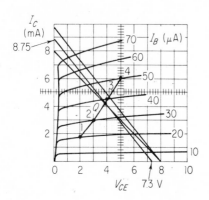

FIG. 10-16.   Self-bias-circuit load and bias lines.

Next, the quiescent operating point must be determined. Usually, this is a very simple calculation. In this circuit, however, there is no logical starting point for finding one of the unknown values, say, I_B or I_C. Every dc value in the circuit is dependent upon the others. For instance, base current depends upon the value of collector voltage, which depends upon the drop across RL, which is a function of collector current, which in turn depends upon base current, etc. Hence, to find where the Q point is located on the load line, we must resort to somewhat devious means. Although the ERCA equations would allow us to find the Q point, an alternative method using the graph itself is shown.

One way of arriving at the operating point is to consider solving simultaneously for two conditions. One of these is collector-current variation for dc conditions, which is already specified by the dc load line. Hence, we need to solve graphically for only one other condition, say, variations in base current.

1.   If V_{CE} were 2 V, base current would be

$$I_B = \frac{V_{CE}}{RB} = \frac{2}{100 \text{ k}\Omega} = 20 \ \mu A$$

This is shown plotted on the graph at point 1.

2. If V_{CE} were 3 V, base current would be

$$I_B = \frac{V_{CE}}{RB} = \frac{3}{100 \text{ k}\Omega} = 30 \ \mu\text{A} \qquad \text{shown as point 2}$$

3. If V_{CE} were 4 V, base current would be

$$I_B = \frac{V_{CE}}{RB} = \frac{4}{100 \text{ k}\Omega} = 40 \ \mu\text{A} \qquad \text{shown as point 3}$$

4. If V_{CE} were 5 V, base current would be

$$I_B = \frac{V_{CE}}{RB} = \frac{5}{100 \text{ k}\Omega} = 50 \ \mu\text{A} \qquad \text{shown as point 4}$$

Connecting these points with a line produces an intersection with the load line. This new line is called the *bias line*, and its intersection with the load line is the Q point.

What we have done above is solve graphically two simultaneous equations. Although there are other ways of arriving at this information, the graphical method is perfectly adequate here, as for most purposes.

From the graph we read that $V_{CE}Q \cong 3.8$ V, while $I_CQ \cong 4.2$ mA. In this instance, then, the dc operating point is not too difficult to obtain.

Large-Signal Analysis

To construct the signal load line we must consider the true ac load, which, to the transistor, appears to be the parallel resistance of RL and r_i.

$$\text{Load}_{\text{ac}} = \frac{RL \times r_i}{RL + r_i} = \frac{1 \text{ k}\Omega \times 5 \text{ k}\Omega}{1 \text{ k}\Omega + 5 \text{ k}\Omega} = 833 \ \Omega$$

Hence, if the collector could rise to V_{CC}, the maximum collector current would be

$$I_{C(\text{max})} = \frac{V_{CC}}{\text{load}_{\text{ac}}} = \frac{8}{833} = 9.60 \text{ mA}$$

The tentative load line is drawn between 9.60 mA on the ordinate to 8 V on the abscissa. The ac load line is therefore drawn parallel to this, but passing through the Q point. From this we determine that $I_{C(max)}$ is about 8.75 mA, while $V_{CE(max)}$ is 7.3 V.

In this special case, we have only an approximate load line because of the feedback from collector to base. In reality, the signal load line for this circuit is even steeper than shown, but for our purposes it is an adequate load line. By modifying the circuit slightly, the load line is accurate, as discussed below.

Small-Signal Analysis

The circuit shown in Fig. 10-15 has many undesirable features because of the negative signal feedback from collector to base. The input resistance is lowered, the voltage gain is greatly reduced, and certain other features are degraded. To overcome these undesirable characteristics, the circuit of Fig. 10-17 is generally used. From a dc standpoint, the two circuits are identical; so the desirable dc feedback that improves the temperature response is retained.

Because C_{F1} has very low reactance at signal frequencies, however, no signal can get from the collector back to the base. The signal gain and input resistance are therefore not affected to any great extent. There is one small effect due to the fact that the total ac load is now shunted by RB_1. RB_1 can usually be made large enough so that it does not significantly add to the total load.

For our purposes, then, the signal load line that has already been constructed is quite accurate enough. Only if RB_1 is less than 10 times the total ac load does it have to be considered. In this case the ac load is 833 Ω. If RB_1 is greater than 8330 Ω, it can be neglected, for most purposes. This is, of course, the present case.

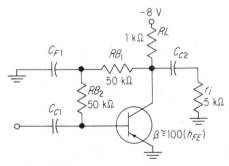

FIG. 10-17.   Self-bias-circuit with only dc feedback.

The voltage gain and input resistance for this circuit are easily determined.

$$r_e = \frac{26}{I_{E\ (mA)}} = \frac{26}{4.2} = 6.2\ \Omega$$

$$r_i = (\beta + 1)r_e = 101 \times 6.2 = 620\ \Omega$$

(Note that RB_2 is also parallel with this, but again, it is so much larger that we can ignore it.)

$$A_v = \frac{\text{load}_{ac}}{r_e} = \frac{833}{6.2} = 134$$

It can be seen that as long as there is no signal feedback, the small-signal conditions are quite straightforward.

10-5 THE UNIVERSAL CIRCUIT WITH COLLECTOR FEEDBACK

The circuit of Fig. 10-18 is similar to the circuit of Fig. 10-17 in that it has collector feedback. Because of this, it is about as difficult to work with, or perhaps a little more so. Again, the use of the ERCA equations would simplify locating the Q point, but the graphical approach will be used here for purposes of illustration.

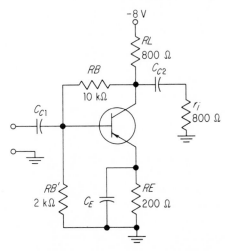

FIG. 10-18.   Example of universal circuit with collector feedback.

DC Analysis

As always, the first step is to construct the dc load line. The dc limits are no different in this circuit than in any other. If the transistor is cut off, V_{CE} is equal to $-V_{CC}$, or 8 V. When the transistor is saturated, $I_{C(max)}$ is

$$I_{C(max)} = \frac{V_{CC}}{RL + RE} = \frac{8}{1\ k\Omega} = 8\ mA$$

A straight line drawn between these two limits, then, is the dc load line for this circuit, as shown in Fig. 10-19.

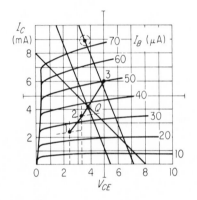

FIG. 10-19. Load and bias lines for universal circuit with collector feedback.

At this point in the analysis it would ordinarily be a simple matter to determine the quiescent operating point. In this circuit, however, it is not quite as simple as was the case in the circuit of Fig. 10-17. The base voltage V_B is determined by the voltage V_C across the voltage divider, which depends upon collector current I_C, which depends upon base current I_B, which depends upon base voltage V_B, etc. In this circuit there is no place to start calculating for the operating point because no value is known. Each value is dependent upon the others, so an indirect approach is called for.

The first step in locating the quiescent operating point for this circuit is to obtain the information necessary to plot the bias line. In this circuit configuration the method is slightly different. Because the voltage across the base-circuit voltage divider is not equal to V_{CE}, we must include external circuit conditions in the determination; from this can be derived the voltage and current values for the transistor itself.

First, several values of the collector voltage referred to ground are assumed. We shall use V_C equal to 3, 4, and 6 V as our starting points. Then emitter current is found, from which base current can be derived. Finally, the value of V_{CE} is determined, and the point is plotted at the proper values of base current and the collector-emitter voltage.

With at least two points plotted, one on each side of the load line, the bias line is drawn, and its intersection with the load line is the operating point. The third point is determined simply to verify, or prove, the bias line. If all three lie upon a slightly curved line, the line is a true one.

The first point is to be found by assuming that V_C is -3 V. With 3 V across RB and RB', the voltage at the junction of the resistors determines V_B, and therefore V_E. Because beta (h_{FE}) is required, read from curves $\beta \cong 100$.

1. Assume $V_C = -3$ V, referred to ground.

$$V_B \cong V_E = \frac{RB'}{RB + RB'} \times V_C = \frac{2 \text{ k}\Omega}{12 \text{ k}\Omega} \times 3 \text{ V} = 0.5 \text{ V}$$

$$I_E = \frac{V_E}{RE} = \frac{0.5 \text{ V}}{200 \text{ }\Omega} = 2.5 \text{ mA}$$

$$I_B = \frac{I_E}{\beta + 1} = \frac{2.5 \text{ mA}}{101} \cong 25 \text{ }\mu\text{A}$$

$$V_{CE} = -V_C + V_E = -3 + 0.5 = -2.5 \text{ V}$$

This information becomes the point labeled 1 on the curves, at $V_{CE} = 2.5$ V and $I_B = 25$ μA.

2. Assume $V_C = -4$ V.

$$V_B \cong V_E = \frac{RB'}{RB + RB'} \times V_C = \frac{2 \text{ k}\Omega}{12 \text{ k}\Omega} \times 4 \text{ V} = 0.667 \text{ V}$$

$$I_E = \frac{V_E}{RE} = \frac{0.667 \text{ V}}{200 \text{ }\Omega} = 3.33 \text{ mA}$$

$$I_B = \frac{I_E}{\beta + 1} = \frac{3.33 \text{ mA}}{101} = 33 \text{ }\mu\text{A}$$

$$V_{CE} = -V_C + V_E = -4 + 0.667 = 3.33 \text{ V} \qquad \text{point 2}$$

3.   Assume $V_C = -6$ V.

$$V_B \cong V_E = \frac{RB'}{RB + RB'} \times V_C = \frac{2\text{ k}\Omega}{12\text{ k}\Omega} \times 6\text{ V} = 1\text{ V}$$

$$I_E = \frac{V_E}{RE} = \frac{1\text{ V}}{200\ \Omega} = 5\text{ mA}$$

$$I_B = \frac{I_E}{\beta + 1} = \frac{5\text{ mA}}{101} = 50\ \mu\text{A}$$

$$V_{CE} = -V_C + V_E = 6 - 1 = 5\text{ V} \qquad \text{point 3}$$

The junction of the bias line and the dc load line is the Q point, as shown, where $V_{CE} = 3.9$ V, and $I_C = 4.1$ mA. Locating the Q point in this circuit, then, is not difficult, although somewhat detailed.

Large-Signal Analysis

The signal load line for this circuit can be constructed by the usual means.

The total ac load is

$$\text{Load}_{\text{ac}} = \frac{RL \times r_i}{RL + r_i} = \frac{0.8\text{ k}\Omega \times 0.8\text{ k}\Omega}{0.8\text{ k}\Omega + 0.8\text{ k}\Omega} = 0.4\text{ k}\Omega$$

Hence a tentative value of $I_{C(\text{max})}$ is

$$I_{C(\text{max})} = \frac{V_{CE(\text{max})}}{\text{load}_{\text{ac}}} = \frac{7.2}{400} = 18\text{ mA} \qquad \left(\begin{array}{l} V_{CE(\text{max})} = V_{CC} - E_{RE} \\ \qquad\quad = 8 - 0.8 = 7.2\text{ V} \end{array} \right)$$

Since 18 mA is not on the graph, we can say that if half of the available voltage appears across the load RL, instead of the full amount of $-V_{CC}$, then half of 18 mA of collector current will flow, or 9 mA.

Therefore, we draw our tentative load line from 7.2 V and $I_C = 0$ to 9 mA and $V_{CE} = V_{CE(\text{max})} - E_{RL} = 7.2 - (7.2/2) = 3.6$ V. This is shown encircled by a dotted line. The true signal load line is run parallel to this, but passing through the Q point. Thus $V_{CE(\text{max})}$ is found to be 5.4 V, while $I_{C(\text{max})}$ is off scale. Its value, however, can be determined as follows:

$$I_{C(\text{max})} = \frac{V_{CE(\text{max})}}{\text{load}_{\text{ac}}} = \frac{5.4}{400} = 13.5\text{ mA}$$

As in the case of the circuit in Fig. 10-15, this is not exactly the true ac load line, because of the negative feedback which tends to lower the value of the ac load somewhat. However, it is quite close enough for present purposes. By slight modification of the circuit, this load line will much more nearly describe this circuit.

Fortunately, the circuit is usually constructed as shown in Fig. 10-20. Again, the circuit possesses dc feedback to help temperature stability, but signal feedback is eliminated because of C_F.

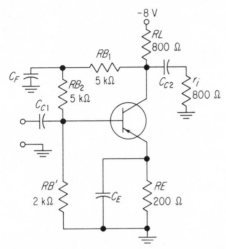

FIG. 10-20.   Circuit arrangement to prevent signal feedback.

The input resistance and voltage gain are affected by this circuit change. The input resistance is now the parallel resistance of RB_2, RB', and the input resistance of the transistor.

The voltage gain is also altered slightly by virtue of the fact that RB_1 is in parallel with the ac load. In both of the foregoing instances (r_i and A_v), the change is very slight, compared with a conventional universal circuit, since the additional resistors are relatively large, which is usually the case. For instance, the ac load of this circuit, not considering RB_1, is 400 Ω. However, if RB_1 is considered, the new value is about 370 Ω, a not-too-significant change.

The input resistance r_i is the parallel resistance of RB', RB_2, and $(\beta + 1)r_e$.

$$R_t = \frac{RB' \times RB_2}{RB' + RB_2} = 1.43 \text{ k}\Omega$$

$$R_{BE} = (\beta + 1)r_e = 101 \times 6.2 = 626 \ \Omega$$

where $r_e = 26/4.2 = 6.2 \ \Omega$

$$r_i = \frac{R_t \times R_{BE}}{R_t + R_{BE}} = \frac{1.43 \ \text{k}\Omega \times 0.626 \ \text{k}\Omega}{1.43 \ \text{k}\Omega + 0.626 \ \text{k}\Omega} = 435 \ \Omega$$

$$A_v = \frac{\text{load}_{ac}}{r_e} = \frac{370}{6.2} = 59.7$$

where $\text{load}_{ac} = \dfrac{1}{1/RL + 1/r_i + 1/RB_1} = 370 \ \Omega$

Thus this circuit, which might appear at first to be quite complex, is actually seen to be rather simple.

10-6   A BIASED-UP DIFFERENTIAL AMPLIFIER

The circuit shown in Fig. 10-21 is commonly referred to as a *biased-up differential-amplifier circuit* because each base is lifted above ground by

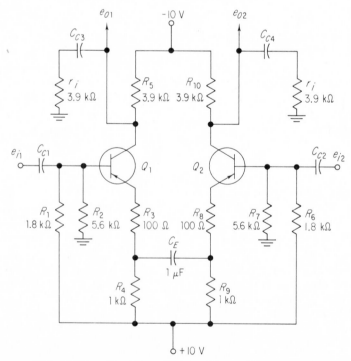

FIG. 10-21.   Biased-up differential-amplifier example.

a voltage divider. The circuit is similar in many respects to both the universal circuit and the dual-supply circuit, as far as circuit biasing is concerned.

A differential amplifier is a circuit that, ideally, amplifies only "differential" (push-pull) signals, rejecting all others. In this particular circuit, the input is a push-pull signal, evident at e_{i1} and e_{i2}. The output is also push-pull, and is seen at e_{o1} and e_{o2}. If either output is referred to the other, rather than ground, the output is said to possess "common-mode rejection." That is, if an unwanted signal appears, for instance, ripple on the $-V_{CC}$ line, the output measured from collector to collector will show little or no ripple. Only valid push-pull signals appearing at the inputs will be amplified.

To analyze the circuit for its operational characteristics, we must first locate the static conditions in the circuit, as always. Considering the circuitry of Q_1, the resistors R_1 and R_2 form the base-biasing voltage divider. The voltage at the base, with reference to ground, is set by these resistors. Likewise, the two resistors R_6 and R_7 form the base-biasing network for Q_2. R_5 is the collector-load resistor for Q_1, and R_{10} serves the same purpose for Q_2. R_3 and R_8 are the emitter-feedback resistors for signal conditions. Resistors R_4 and R_9 are instrumental in helping to set the emitter current consistent with the available supply voltage. The two capacitors, C_{c1} and C_{c2}, are the input-coupling capacitors, and C_E is the emitter-bypass capacitor. C_{c3} and C_{c4} are the output-coupling capacitors, and r_i represents the input resistance of the next stage.

There are several ways to analyze such a circuit. We shall follow the graphical approach for both dc and signal analysis, using the ERCA equations where applicable only to verify the results obtained graphically.

DC Analysis

Since both transistors are operating in the same circuit, we can use either for the example. The curves for either of the two transistors are shown in Fig. 10-22. The dc load line for one transistor is constructed between the two extreme static limits of circuit operation. With the transistor cut off, $V_{CE(\text{max})}$ is 20 V. This is one limit of the dc load line. The other limit is attained with the transistor in saturation, when $I_{C(\text{max})}$ is determined by the total applied voltage and the total resistance.

$$I_{C\text{max}} = \frac{|V_{\text{total}}|}{R_{\text{total}}} = \frac{|20|}{3.9 \text{ k}\Omega + 1.1 \text{ k}\Omega} = 4 \text{ mA}$$

A straight line drawn between these two limits is the dc load line, as shown. On the assumption that I_C and I_E are nearly the same, we can now locate the quiescent operating point on the curve.

The voltage at the base, with reference to ground, is easily determined. This will allow us to find the collector current, which will specify the Q point.

$$V_B = \frac{R_2}{R_1 + R_2} \times V_{EE} = \frac{5.6 \text{ k}\Omega}{1.8 \text{ k}\Omega + 5.6 \text{ k}\Omega} \times 10 \cong 7.6 \text{ V}$$

By assuming a 0.1-V drop across the forward-biased junction, we can accurately determine emitter current.

$$I_E = \frac{V_{EE} - (V_B + 0.1)}{R_{E\text{(total)}}} = \frac{10 - (7.6 + 0.1)}{1.1 \text{ k}\Omega} = \frac{2.3}{1.1 \text{ k}\Omega} \cong 2.1 \text{ mA}$$

$$E_{RL} = I_C \times R_L = 2.1 \text{ mA} \times 3.9 \text{ k}\Omega \cong 8.2 \text{ V} \qquad I_C \cong I_E$$

The junction of the dc load line and $I_C = 2.1$ mA is the Q point.

Because the circuit is symmetrical, these values are true for either transistor. Note that the circuit has been designed for about equal

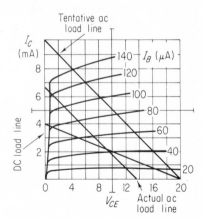

FIG. 10-22. Load lines for differential amplifier.

voltage drops across both RL and the transistor. The values of dc voltage around the circuit may now be calculated. From the dc load line, V_{CE} is about 9.5 V. Or, this may be determined as follows. First, find E_{RE}.

$$E_{RE} = V_{EE} - (V_B + 0.1) = 10 - 7.7 = 2.3 \text{ V}$$

$$V_{CE} = V_{\text{total}} - E_{RL} - E_{RE} = 20 - 8.2 - 2.3$$

$$= 9.5 \text{ V}$$

V_C, the collector voltage referred to ground, is calculated by equating the collector circuit voltage drops for this value.

$$V_C = -V_{CC} + E_{RL} = -10 + 8.2 = -1.8 \text{ V}$$

An alternative method of arriving at the dc conditions in the circuit is to use the ERCA equations. This is somewhat more straightforward and less detailed than the more conventional approach. However, because the circuit is not quite the same as the conventional universal circuit, the ERCA equation is slightly different.

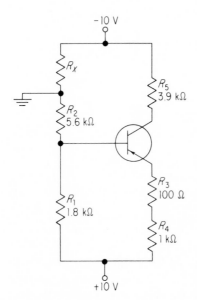

FIG. 10-23. Circuit arrangement to derive R_X for differential amplifier.

To modify the equation for use with a circuit such as this, an imaginary resistor, R_X, is used to complete the bridge circuit, as shown in Fig. 10-23. Since the junction of R_X and R_2 is always at ground, the value of R_X (or rather, the value it would be if it existed), to place its lower end at ground, is easily found.

$$R_X = \frac{V_{CC}(R_1 + R_2)}{V_{EE}} = \frac{10 \times 7.4 \text{ k}\Omega}{10} = 7.4 \text{ k}\Omega$$

The value of R_X plus R_2 becomes the resistor that we usually call RB in the usual ERCA equation, and here we call it R_{BX}.

$$R_{BX} = 7.4 \text{ k}\Omega + 5.6 \text{ k}\Omega = 13 \text{ k}\Omega$$

Now the circuit can be analyzed in the same way as any universal circuit yet encountered, substituting R_{BX} in place of RB in the equation. Also, R_1 is RB', R_5 is RL, and $R_3 + R_4$ is RE, of course.

$$R_{CB} = \frac{(R_{BX})(RE)}{R_{B'}} - RL = \frac{13 \text{ k}\Omega \times 1.1 \text{ k}\Omega}{1.8 \text{ k}\Omega} - 3.9 \text{ k}\Omega$$

$$= 4.04 \text{ k}\Omega \cong 4 \text{ k}\Omega$$

From this point, the dc analysis is identical with any other.

$$\gamma = \frac{R_{CB}}{R_{CB} + RL + RE} = \frac{4.0 \text{ k}\Omega}{4.0 \text{ k}\Omega + 3.9 \text{ k}\Omega + 1.1 \text{ k}\Omega} = 0.444$$

$$V_{CE} = \frac{R_{CB}}{R_{CB} + RL + RE} \times V_{\text{total}} = 8.89 \text{ V}$$

$$I_C = \frac{V_{CE}}{R_{CB}} = \frac{8.89}{4 \text{ k}\Omega} = 2.2 \text{ mA}$$

Thus the modified ERCA equation will allow swift and simple analysis for the dc state of the circuit. The two methods of dc analysis are seen to agree very closely.

Large-Signal Analysis

For the circuit in Fig. 10-21, the signal load line for either transistor depends on the total value of the ac load. Using Q_1 as an example, the signal load line will be constructed for the given circuit values. This is done in exactly the same manner as in preceding circuit examples. The two tentative circuit limits are found, and the tentative load line is drawn between these limits.

If the transistor is cut off, $V_{CE(max)}$ is equal to $V_{CC(total)}$, or 20 V. If the transistor is just saturated, the total ac load consists of the parallel combination of RL and r_i in series with the unbypassed emitter resistor.

$$\text{Load}_{ac} = \frac{R_5 \times r_i}{R_5 + r_i} = \frac{3.9 \text{ k}\Omega \times 3.9 \text{ k}\Omega}{3.9 \text{ k}\Omega + 3.9 \text{ k}\Omega} = 1.95 \text{ k}\Omega$$

This, plus R_3, yields a value of 2.05 kΩ, or approximately 2 kΩ. Thus maximum collector current, if $V_{CE(\text{max})} = V_{CC}$, would be

$$I_{C(\text{max})} = \frac{V_{CC}}{\text{load}_{ac(\text{total})}} = \frac{20}{2 \text{ k}\Omega} = 10 \text{ mA}$$

The tentative load line is drawn between these two limits, as shown in Fig. 10-22. This line has the correct slope, but, as before, does not pass through the Q point. A line passing through the Q point and parallel to the tentative load line is the true ac load line.

From the actual load line the values of $V_{CE(\text{max})}$ and $I_{C(\text{max})}$ can be derived. $V_{CE(\text{max})}$ is seen to be 13.3 V, while $I_{C(\text{max})}$ is about 6.7 mA.

Small-Signal Analysis

The small-signal characteristics of this circuit that concern us are the voltage gain and the input resistance. The voltage gain of this circuit is a function of the ac load and the unbypassed emitter resistance, assuming r_e to be negligible compared with 100 Ω.

$$A_v = \frac{\text{load}_{ac}}{R_3} = \frac{1.95 \text{ k}\Omega}{100} = 19.5$$

The input resistance of Q_1 is essentially the parallel resistance of R_1, R_2, and the input resistance of the transistor. (Assume $\beta = h_{fe} = 100$.)

$$R_P = \frac{R_1 \times R_2}{R_1 + R_2} = \frac{1.8 \text{ k}\Omega \times 5.6 \text{ k}\Omega}{1.8 \text{ k}\Omega + 5.6 \text{ k}\Omega} = 1.36 \text{ k}\Omega$$

$$R_{ib} = (\beta + 1)(r_e + R_E) = (101)(13 + 100) = 11.4 \text{ k}\Omega$$

where $r_e = 26/I_E = 26/2 = 13$ Ω

$$r_i = \frac{R_P \times R_{ib}}{R_P + R_{ib}} = \frac{1.36 \text{ k}\Omega \times 11.4 \text{ k}\Omega}{1.36 \text{ k}\Omega + 11.4 \text{ k}\Omega} = 1.22 \text{ k}\Omega$$

Thus the small-signal characteristics of interest are easily determined.

QUESTIONS AND PROBLEMS

Because the answers to these numerical problems are subject to the normal variations encountered in reading the graph, allow some leeway in your answers when comparing them with those given in the Answers section at the end of the book. In other words, if your answer is not exactly the same, but is close, consider it correct.

10-1 Refer to Fig. 10-1. Using the characteristic curves given in Fig. 10-2, construct the dc and ac load lines if $-V_{CC}$ is changed to -8 V. For signal conditions, what is $V_{CE(max)}$? $I_{C(max)}$?

10-2 Refer to Fig. 10-13. Using the characteristic curves given in Fig. 10-14, construct the dc and ac load lines if $-V_{CC} = -6$ V and $V_{EE} = +3$ V. What are $V_{CE(max)}$ and $I_{C(max)}$ for signal conditions?

10-3 Refer to Fig. 10-18. Discuss the effect on the ac load line of changing the value of RB' to a lower value.

10-4 Refer to Fig. 10-10. Discuss the causes of I_1 and I_2.

10-5 Refer to Fig. 10-9. Discuss the reason for $I_{RL(min)}$.

10-6 Discuss why the ac load line is always steeper than the dc load line.

10-7 Discuss your reasons for the load line (ac) always passing through the dc quiescent point.

10-8 Refer to Figs. 10-1 and 10-2. Change only V_{CC} to -10 V. Determine graphically the static (dc) values of $V_{CE(max)}$ and $I_{C(max)}$; the dynamic (ac) values of $V_{CE(max)}$ and $I_{C(max)}$.

10-9 Refer to Figs. 10-6 and 10-7. Change only r_i to 3000 Ω. Graphically determine the values of static $V_{CE(max)}$ and $I_{C(max)}$; the values of dynamic $V_{CE(max)}$ and $I_{C(max)}$.

10-10 Refer to Fig. 10-6. The resistor shown as r_i is to be changed from 4000 to 7000 Ω. Find the new value of I_{min}.

10-11   Refer to Question 10-10. Find the new value of the sum of I_1 and I_2 (I_{Ctotal}).

10-12   Refer to Fig. 10-18. Change only RL from 800 to 1200 Ω. Graphically analyze the circuit for the quiescent values of V_{CE} and I_C.

10-13   Refer to Fig. 10-17. Briefly discuss the circuit action if C_{F1} is removed from the circuit.

10-14   Refer to Fig. 10-13. Change r_i from 3700 to 2200 Ω. Graphically solve for the midband ac limits $V_{CE(max)}$ and $I_{C(max)}$. Assume $V_{BE} = 0.25$ V and $h_{FE} = 100$.

10-15   Refer to Fig. 10-13. Change only V_{EE} from 6 to 12 V. Graphically analyze for the static value of $V_{CE(Q)}$ and $I_{C(Q)}$; determine the circuit gamma.

10-16   Refer to Fig. 10-15. Change RB from 100,000 to 75,000 Ω. Graphically solve for the quiescent operating point as specified by $V_{CE(Q)}$.

CHAPTER 11 TEMPERATURE CHARACTERISTICS

11-1 BIAS DEFINED

The need for relatively complex bias circuits arises because all transistors are temperature-sensitive to a greater or lesser degree. Of the two types of transistors, germanium and silicon, germanium is by far more sensitive to temperature variations. Therefore much of what we have to say about temperature characteristics of transistor circuits concerns those using germanium transistors. Germanium transistors typically have leakage currents in the range of 1 to 10 μA or more, while silicon transistors often have leakage currents in the range of 1 to 100 nA (nanoamperes). Thus leakage current is seldom as great a problem when using silicon transistors, but the transistor data sheet should be the final authority.

The simpler circuits used to bias a germanium transistor will work satisfactorily only over a very narrow temperature range. Because of this, the biasing circuits become more and more complex, to compensate for the maximum temperature at which the circuit is expected to perform.

The main purpose of the biasing network of a linear amplifier is to prepare the transistor for the job of amplifying. That is, with no signal applied to the circuit, the transistor must be turned on to some extent. Some amount of collector current, as well as base and emitter current, must be flowing. The amount of these currents must be intermediate between the maximum and minimum allowed by the external circuit. Expressed still another way, if the transistor is to operate as a linear amplifier, it must be biased so that its emitter-base junction is forward-biased and its collector-base junction is reverse-biased. This condition must always exist under all conditions of dc and signal operation if linear amplification is to be achieved.

The secondary purpose of the biasing circuitry is to prevent a shift in operating point, or at least minimize it, as the temperature increases. In some cases the biasing arrangement does its job admirably, but in others it fails badly. In order to perform both these functions well, the bias circuits are somewhat complex, as we shall see. First, however, we shall want to know more about the leakage current I_{cbo} that causes this change in the operating point as the temperature changes.

11-2 LEAKAGE CURRENT AND DELTA V_{BE}

Recalling from basic transistor theory that the leakage current I_{cbo} is caused by thermal agitation, we know that it is the breaking of the covalent bonds that produces the electron-hole pairs. If the covalent bond is broken in the vicinity of the reverse-biased collector junction, a carrier will be produced that can contribute to current flow in the collector circuit. Figure 11-1a shows this diagrammatically, and the

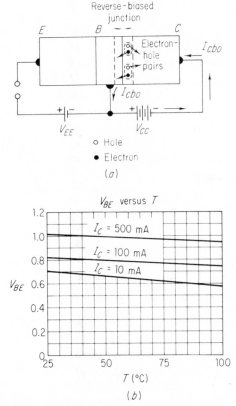

FIG. 11-1. (a) Illustrating I_{cbo} generation. (b) V_{BE} change versus temperature.

generation of electron-hole pairs in the collector barrier region is clearly seen. We should usually consider that there are no free holes, or electrons, in this region, but because of thermal activity, they exist in a number that is directly proportional to the temperature of the material. I_{cbo} is normally measured with the emitter open, as shown. A microammeter in either the collector or the base lead would measure the amount of leakage current for some given temperature.

Leakage current approximately doubles for every 10°C rise in temperature in the case of germanium, and can become a very large amount of current at the higher temperatures. For instance, if I_{cbo} for a particular transistor is 4 μA at room temperature (25°C), it will become about 65 μA at 75°C. In some circuits, this could significantly alter the dc operating point. (I_{cbo} in silicon transistors increases 2 times for every 6°C.)

An additional factor that influences the temperature response of a transistor is the variation of V_{BE} with a temperature change. Base-emitter voltage *decreases* as the temperature increases, and in many circuits this further alters the quiescent operating point.

Figure 11-1b details the relationship between V_{BE} and temperature for a typical transistor. From the curves, the change in V_{BE} per degree Celsius can be determined as being 1.6 mV/°C. That is, for every degree of temperature rise, V_{BE} decreases 0.0016 V.

In the beta-dependent circuits, this effect adds further to the instability of the circuit. However, in a beta-independent circuit the effect is almost negligible. In the beta-dependent case, a decrease in V_{BE} causes an increase in base current, which is therefore amplified. In the more stable configurations, emitter current is slightly increased, which is, of course, not amplified. The amount of current increase is on the order of a fraction of a microampere to a few microamperes per degree Celsius, and is determined to a large degree by the external circuit components.

11-3 TEMPERATURE-CHANGE EXAMPLE

In order to see how a change in temperature affects a typical circuit, let us investigate the circuit of Fig. 11-2. Then, when we have an understanding of what is going on in the circuit, we can further investigate some of the ideas of temperature stability and biasing. The circuit shown is a simple common-emitter amplifier, and it appears along with its characteristic curves. Constructed on the curves is the dc load line for the circuit, and the quiescent operating point is shown as point Q. The quiescent operating point is chosen by the circuit designer to fulfill the requirements of the amplifying job to be done by the circuit. Its position on the load line is determined by the characteristics of the transistor and the values of *RL* and *RB*.

The normal operating point is chosen to exist at room temperature. The values of the components are picked to accomplish this,

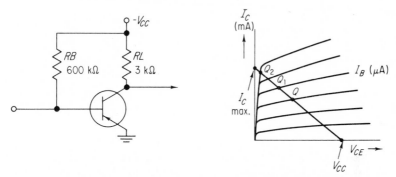

FIG. 11-2. *Q*-point drift due to temperature change.

as well as to help maintain this *Q* point as the temperature varies. In this simple circuit, however, the drift of the operating point is impossible to eliminate altogether, and very little can be done to stabilize it for temperature variations.

Let us see how a change in temperature alters the normal quiescent operating point of this circuit. At normal temperatures, the amount of I_{cbo} is negligible in most cases; so it does not greatly influence the operating point. Thus the operating point as shown on the curves is at point *Q*.

If the temperature should increase for any reason, I_{cbo} increases by a large amount, about doubling for every 10°C increase. As the device becomes hotter, the total collector current increases. This increase is due only to the temperature rise.

As more collector current flows because of increased I_{cbo}, the transistor is turned on harder, and the operating point moves to, perhaps, point Q_1. The transistor is now operating with increased collector, base, and emitter current, but with reduced collector-to-emitter voltage.

If the temperature rises still further, the transistor turns on even harder, and could easily go to point Q_2. Now the transistor is saturated, and is not capable of amplifying. V_{CE} is minimum, perhaps 0.1 V, while I_C has increased to the very maximum that the circuit constants will allow.

Thus, with this simple circuit, the operating point on the load line has shifted from its normal position to saturation, with only a moderate increase in temperature. For a typical transistor in a circuit such as that shown in Fig. 11-2, this could occur for a temperature increase of as little as 20 or 30°F.

11-4 TEMPERATURE STABILITY OF THE *CC, CE,* AND *CB* CIRCUITS

Each of the three basic circuit configurations reacts differently to an increase in temperature. Of the three, the common emitter is the most sensitive to an increase in temperature, the other two circuits being far more stable as far as temperature variations are concerned. In order to understand why these differences occur, we should investigate the action of I_{cbo} in the transistor circuit.

Figure 11-3 shows a simplified drawing of a transistor, indicating the direction and path of I_{cbo} as it is usually measured. With the emitter open, the leakage current has no choice but to travel from

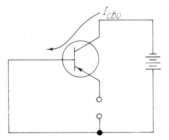
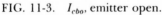
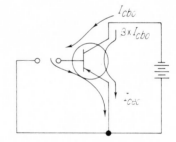

FIG. 11-3. I_{cbo}, emitter open.   FIG. 11-4. I_{cbo}, base open.

the collector to the base in the direction indicated. The same amount of I_{cbo} flows in both the collector and the base lead. In some circuit arrangements, this is the path and direction that the leakage current takes.

Figure 11-4 shows the other extreme. Here the base lead is open, rather than the emitter lead, and now I_{cbo} must flow across the base region into the emitter, as shown. The fact that current is flowing from the base into the emitter means that the transistor will amplify this current. That is, the transistor cannot distinguish between normal base current and this leakage current, and $\beta \times I_{cbo}$ will also flow in the collector circuit. Thus not only does I_{cbo} itself flow in the collector circuit, but also *beta times* I_{cbo} flows. Therefore the true expression for total collector leakage current I_{cco} in this configuration is as follows:

$$I_{C(\min)} = I_{cbo} + \beta I_{cbo} = (\beta + 1)I_{cbo} = I_{ceo}$$

Many practical circuits lie somewhere between these two extremes. That is, a portion of I_{cbo} is amplified by the transistor, and a portion of it is not. It is therefore not a simple matter to analyze a circuit for its exact temperature characteristics.

A few general rules can be formulated to specify the way the three basic circuit configurations react to changes in temperature. Generally speaking, the common-base circuit usually resembles the circuit of Fig. 11-3, and so it does not amplify I_{cbo}, since the leakage current flows out the base lead and into the external circuit. On the other hand, the common-emitter circuit usually resembles the circuit arrangement of Fig. 11-4. Thus the total leakage current flowing in the collector circuit is many times that value that would be measured in the typical test circuit of Fig. 11-3. For this reason, the basic common-emitter circuit has the worst possible temperature characteristics of any of the three configurations.

Somewhere between these two circuits lies the emitter follower or common-collector circuit, and the temperature stability of this circuit is dependent upon the circuit constants to a large degree. It can be made to be as good as the common-base circuit, or it can be made to be as poor as the common-emitter circuit. The external circuit values are the determining factor.

11-5 STABILITY FACTOR S

One of the commonly accepted ways of determining whether a particular circuit has good or poor thermal stability is by the use of the so-called stability factor S. The stability factor of a circuit may be defined as the ratio of the *change* in collector current to the *change* in reverse saturation current, I_{cbo}, which caused the collector current to change. The equation for this relation is

$$S = \frac{\Delta I_C}{\Delta I_{cbo}}$$

The stability factor of a circuit tells us what effect a change in I_{cbo} has upon the collector current. In practice, low values of S are desirable, for this indicates a stable circuit. The lowest possible is unity. On the other hand, if the S factor for a particular circuit is high, this indicates that the thermal stability is very poor. The highest value of S for a given case is the dc beta, or h_{FE}, of the transistor itself, approximately.

For a germanium transistor in a typical circuit, a value of S between 3 and 5 is found to be adequate for a temperature rise of about 80°C.

There is no hard and fast rule for using the S factor. Each circuit must be evaluated on its own merits.

A formula that will allow one to determine the S factor for any circuit is

$$S = \frac{1 + RB/RE}{1 + (1 - \alpha)(RB/RE)}$$

where RB = total series-parallel resistance in the base
RE = total series dc resistance in the emitter
α = dc alpha of the transistor

From this formula it can be seen that the larger values of resistance in the emitter lead tend to decrease the S factor, while large values of resistance in the base lead tend to increase the S factor.

As an example, let us work out the S factor of the circuit of Fig. 11-2. Assume β_{dc} is 50, and assume the resistance of the emitter itself is 26 Ω ($-V_{CC} = -12$ V).

$$S = \frac{1 + RB/RE}{1 + (1 - \alpha)(RB/RE)} = \frac{1 + (600 \text{ k}\Omega/26)}{1 + (1 - 0.98)(600 \text{ k}\Omega/26)}$$

$$= \frac{1 + 23,000}{1 + (0.02 \times 23,000)} = \frac{23,001}{462} = 49.8$$

Thus the S factor is nearly equal to β_{dc}, and the circuit is doing nothing to prevent the drift in the operating point as the temperature changes.

The S factor, then, is used simply to determine the *relative* merits of a particular circuit with regard to its thermal stability. As the S factor approaches 1, the thermal characteristics are good. But the thermal characteristics are poor if the S factor approaches the dc beta of the transistor.

11-6 THE SEVEN BASIC BIASING CIRCUITS

The following seven basic circuits (circuits a through g), which were discussed earlier, are shown in the CE configuration, and will cover about 95 percent of all the transistor linear-amplifier circuits one might expect to encounter.

Fixed Bias — Circuit a

Circuit a, Fig. 11-5, is the same as that in Fig. 11-2. The others are shown in an order that becomes increasingly more complex. As we

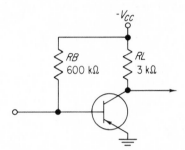

FIG. 11-5. Fixed-bias circuit.

shall see in the general case, the more complex the circuit, the better its temperature stability, and the better the general characteristics of the biasing arrangements.

In Sec. 11-5, we determined the S factor for circuit a, and we discovered that it is not at all stable for changes in temperature. We previously found that this circuit is beta-dependent, and now we find that it is also temperature-dependent. Thus it does a poor job of providing a stable and proper quiescent operating point.

The quiescent operating conditions for this circuit configuration are completely dependent upon the characteristics of the transistor itself. If a new transistor is substituted for the old one, the operating point on the load line may be quite different. Thus it is nearly impossible to bias this circuit properly without handpicking the component parts. In a practical circuit, the base resistor is often made to be a potentiometer, so that differences in transistor characteristics can be accommodated when the transistor is changed. For this reason, and because of its poor thermal stability as well, the circuit is seldom seen in practical usage.

Fixed Bias with Emitter Resistance — Circuit b

In Fig. 11-6 we show a slightly different circuit. Here we have added an emitter resistance, and the effect of this is to lower the S factor and thus provide the circuit with better temperature stability. Just how the added resistor helps to stabilize the temperature response is best explained by comparing it with the circuit of Fig. 11-5.

If the emitter is at ground, as in Fig. 11-5, the circuit is most sensitive to temperature changes. The emitter can be at no other potential but ground, and as the temperature increases, I_{cbo} increases. Therefore the emitter current increases too. As the emitter current increases and flows from the base to the emitter across the junction, it encounters a certain resistance, R_{EE}, and so creates a voltage drop. This drop is in such a direction as to increase the forward bias on the junction slightly by making the base more negative (PNP case). Thus the transistor is turned on harder, causing more current to flow,

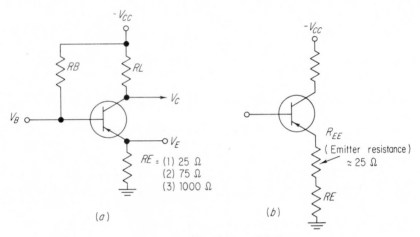

FIG. 11-6.   Fixed-bias with RE.

which produces more heat and more I_{cbo}, etc. As the current increases, the base is driven in a direction to turn the transistor on more, until finally the transistor is driven to saturation, or destroyed if collector current is not limited to a safe value by external resistors.

Now, if an emitter resistor is used, as in Fig. 11-6, and if its value is 25 Ω (the same as the emitter resistance), the change of voltage from base to emitter is only half as much as it would be if the emitter were directly grounded. This is true because half of the total voltage created by the increased current is dropped across the emitter resistance. This results in a smaller increase from base to emitter, and the transistor is not turned on as hard as if RE were not present.

If RE is made to be 75 Ω, only $^{25}/_{100}$, or $\frac{1}{4}$, of the total voltage is developed across the transistor emitter resistance, and the base is turned on only one-fourth as hard as if RE were not present. When RE is made much larger, the increase in the drop across the emitter-base junction due to an increase in temperature is negligible. Thus the transistor does not tend to turn on harder to any appreciable extent. The added resistor has greatly increased the circuit's ability to withstand a temperature increase.

Self-Bias—Circuit c

The self-bias circuit, in Fig. 11-7, is quite different from the others shown so far in this chapter. The base resistor being returned to the collector instead of the power supply causes numerous changes, not the least of which is improved temperature stability.

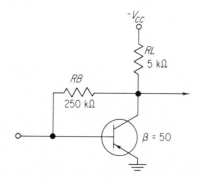

FIG. 11-7. Self-bias circuit.

This circuit has a self-adjusting feature that tries to turn the transistor off if collector current tries to increase because of a temperature variation. The operation of this self-adjusting feature is not difficult to understand. If collector current increases, this causes a larger voltage drop across the load resistor RL, and V_C falls toward ground. With V_C more positive than it was, there is a smaller voltage across RB, and therefore less current will flow through it. The current through RB is the base current, and less base current tends to turn off the transistor somewhat. This, of course, causes V_C to return to the original value to some degree.

The net result of this action is to compensate somewhat for any increase in I_{cbo} by turning the transistor partly off as I_{cbo} increases. The voltage at V_C, then, tends to remain quite constant over a wide range of temperature variation.

In this circuit, RB is often two separate resistors in series. The junction between these two resistors is bypassed to ground with a large capacitor to reduce the signal, or ac, feedback. Thus the signal voltage gain is not reduced. Since the changes due to temperature variations occur slowly, the capacitor does not alter the dc condition at all. In this modified form, the circuit has very good thermal characteristics, as well as high voltage gain.

Self-Bias with Emitter Resistor—Circuit *d*

This circuit, shown in Fig. 11-8, has the temperature attributes of the two preceding circuits. Both collector and emitter feedback for dc

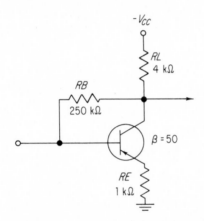

FIG. 11-8. Self-bias with emitter resistance.

conditions are evident, and this gives rise to even better temperature stability. The descriptions just given for circuits *b* and *c* are valid for this circuit and will not be repeated here.

The Voltage Divider—Circuit *e* (Universal Circuit)

The so-called universal circuit is shown in Fig. 11-9. This circuit is quite different from any of the previous four circuits in that the base

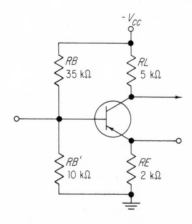

FIG. 11-9. The universal circuit.

voltage is set, not by any transistor characteristic, but by the voltage divider in the base circuit. This circuit, with properly chosen components, is by far the most desirable of any yet presented. It can be designed to be virtually independent of transistor characteristics as far as the quiescent operating point is concerned. Also, its temperature stability is excellent, compared with the previously shown circuits.

The large resistor in the emitter gives added temperature stability in the same way as for the preceding circuits. The voltage divider is an added help in this respect, also keeping the operating point constant over a wide range of temperature variations. To understand how this can be, consider that the normal base current is usually on the order of 50 μA. If resistors RB and RB' are properly chosen, the current through them can be made much larger than the base current, say, 500 μA. In this case, the voltage at the junction of these two resistors is primarily a function of the resistor current, rather than the base current. Now, if base current should increase, the voltage at this junction, V_B, must remain relatively constant, since even if I_B doubles, the increase does not produce a very significant voltage drop across these resistors. This is true as long as the normal current in the resistors is many times that of the base current itself.

Also, the addition of RB' gives a path for a portion of I_{cbo} to flow out the base lead to ground. That amount of I_{cbo} that does not flow across the base-emitter junction will not be amplified by the transistor. Thus the ill effects of the leakage current are further reduced. The smaller RB' can be made, the better the temperature stability will be.

It is instructive to calculate the stability factor for this circuit and compare it with the S factor of the circuit in Fig. 11-5, which we found to be about 50. ($\beta = 50$, $\alpha = 0.98$.)

$$S = \frac{1 + RB/RE}{1 + (1 - \alpha)(RB/RE)} = \frac{1 + (7.78 \text{ k}\Omega/2 \text{ k}\Omega)}{1 + (1 - 0.98)(7.78 \text{ k}\Omega/2 \text{ k}\Omega)}$$

$$= \frac{1 + 3.89}{1 + (0.02 \times 3.89)} = 4.54$$

where $RB = \dfrac{35 \text{ k}\Omega \times 10 \text{ k}\Omega}{35 \text{ k}\Omega + 10 \text{ k}\Omega} = 7.78 \text{ k}\Omega$

$RE = 2 \text{ k}\Omega$

Thus with an S factor of 4.54 it is seen that the thermal stability of this circuit is very good.

The Universal Circuit with Collector Feedback—Circuit f

This circuit, shown in Fig. 11-10, is quite similar to circuit e. The base resistor RB, being returned to the collector, adds still further to the temperature stability by the same mechanism as described for the circuit in Fig. 11-7. The reasons given previously are just as applicable in this case.

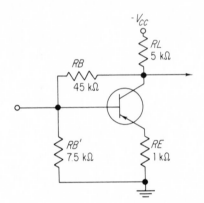

FIG. 11-10. Universal circuit with collector feedback.

The Dual-Supply Bias—Circuit g

The circuit shown in Fig. 11-11a is very closely related to the universal circuit shown in Fig. 11-9. With properly chosen components, this circuit can be made to have the best temperature stability of any shown. Also, its dependence upon the transistor characteristics is very slight, as far as the dc operating point is concerned. These factors cause this circuit to be a very desirable configuration. Also, the circuit is easily adaptable to any of the three basic configurations. It can be used as a CE, CC, or CB amplifier with only slight modification.

To understand why this circuit is related to the universal circuit, we must develop it as shown in Fig. 11-11b and c. In Fig. 11-11b, the regular universal circuit is shown. The base voltage must be some value between −10 and +10 V. One of these values is ground (0 V), and the voltage divider must divide the available voltage so that the junction between the two resistors is at ground. But why use two resistors to obtain ground when it is already available? One need simply connect the base to ground through a suitable resistor (Fig. 11-11c) and eliminate the other one. Thus the circuit is seen to be very similar to the universal circuit when it is to be used with two power supplies.

In this circuit, the base resistor does not determine the amount of base current, which is simply a function of the transistor's characteristics, being equal to $I_E/(\beta + 1)$.

As stated earlier, this circuit has excellent temperature stability. Let us calculate the S factor for the circuit of Fig. 11-11a in order to verify this statement. (Assume $\alpha = 0.98$.)

$$S = \frac{1 + RB/RE}{1 + (1 - \alpha)(RB/RE)} = \frac{1 + (4.7 \text{ k}\Omega/2.2 \text{ k}\Omega)}{1 + (0.02)(4.27 \text{ k}\Omega/2.2 \text{ k}\Omega)} = 3.0$$

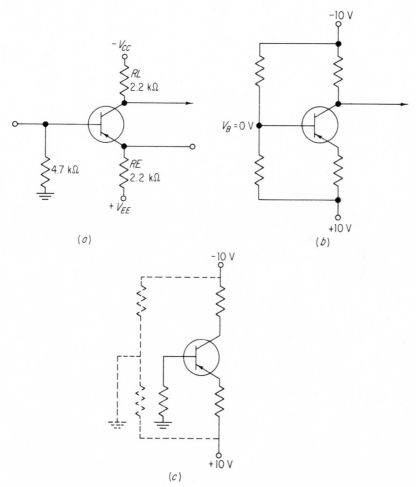

FIG. 11-11. Dual-supply universal circuit.

The S factor approaches unity (1), and so the stability is especially good. (If S equals 1, temperature stability is the best possible, and temperature has no effect on the circuit up to the maximum for the basic material. For germanium this is 185°F, and for silicon this is 300°F.)

11-7 MAXIMUM-POWER-DISSIPATION CURVE

Closely associated with the temperature characteristics of a transistor is its maximum collector power dissipation. When operating, any transistor can function with no more than its maximum power rating being dissipated. The collector power is simply the product of I_C and V_{CE}.

As an example, if V_{CE} is 2 V and I_C is 160 mA, the collector dissipation is

$$P_C = V_{CE} \times I_C = 2 \times 0.16 = 0.32 \text{ W, or } 320 \text{ mW}$$

The manufacturer usually gives the maximum collector dissipation as some number of watts, or milliwatts, of power dissipation in free air at 25°C. If the transistor is expected to be operated at a higher temperature than this, it must be derated, and thus can dissipate less power at this elevated temperature. For example, a typical transistor is rated at 200 mW at 25°C. The manufacturer claims that it must be derated 2.67 mW/°C if the temperature is greater than 25°C. If the transistor is to be operated at a temperature of 55°C, it must be derated by 30 × 2.67, or 80 mW. Therefore, at 55°C it must be allowed to dissipate no more than 200 − 80, or 120 mW. The external circuit must be designed so that no more than 120 mW is dissipated at the collector junction.

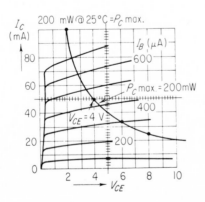

FIG. 11-12. Maximum collector-power dissipation curve.

The maximum-power-dissipation curve can be plotted very conveniently on the collector curves of the transistor. By doing this, the load line can easily be kept below the power curve to ensure that the transistor will at no time exceed its maximum rating. Such a curve is shown in Fig. 11-12. The points generating the curve are determined as follows. First choose several arbitrary values of V_{CE}. Then calculate the current necessary to produce $P_{C(max)}$.

$$P_{C(max)} = I_C \times V_{CE}$$

Therefore

$$I_C = \frac{P_{C(max)}}{V_{CE}}$$

If $P_{C(max)} = 0.2$ W and $V_{CE} = 2$ V,

$$I_C = \frac{0.2}{2} = 0.1 \text{ A}$$

If $P_{C(max)} = 0.2$ W and $V_{CE} = 4$ V,

$$I_C = \frac{0.2}{4} = 0.05 \text{ A}$$

If $P_{C(max)} = 0.2$ W and $V_{CE} = 6$ V,

$$I_C = \frac{0.2}{6} = 0.033 \text{ A}$$

If $P_{C(max)} = 0.2$ W and $V_{CE} = 8$ V,

$$I_C = \frac{0.2}{8} = 0.025 \text{ A}$$

These points, plotted on the curves at the proper values of V_{CE} and I_C, when connected together, form the curve of maximum power dissipation. Operation above the curve *for any length of time* is forbidden, and normal operation must be kept below it to ensure less than maximum dissipation.

11-8 TEMPERATURE ANALYSIS

The foregoing discussion contains the basic information necessary to understand the problems pertaining to the thermal stability of transistor circuits. The next step is to find a way of actually calculating the values of a circuit and of determining the temperature at which it can be expected to fail. In order to do this, one must be able to specify completely the dc response of any circuit. The quiescent values, and the circuit limits, must be known before any quantitative discussion of temperature response can be started.

One of the problems in determining the temperature response of a circuit is that it is difficult to obtain exact data on a particular transistor with regard to its temperature response. By making several simplifying assumptions we can very closely approximate the temperature at which nearly any basic circuit will begin to fail under worst-case conditions.

One of these assumptions is that, for germanium transistors, I_{cbo} will double for every 10°C rise in temperature. For silicon, we assume that I_{cbo} will double for every 6°C rise. This may or may not be exactly true for a given transistor, but it will give very satisfactory results, as we shall see. Also, we assume the transistor's beta remains constant over the allowable temperature range, but this too is not exactly true. Again, however, since we are looking for worst-case conditions, this does not greatly influence our results.

Generally, the transistor manufacturer gives I_{cbo} at the greatest V_{CE} for which the transistor is rated. This is a worst-case condition, and so will yield results that are worst-case also. For example, a circuit that is analyzed to fail at 150°C will, actually, very probably fail at some higher temperature. If the circuit does work above the calculated figure, so much the better. We want to know only the lowest possible maximum temperature that will allow the transistor to still function properly.

The first step in analyzing the circuit for its temperature response is to decide at what point the transistor can be considered to have failed. This depends, in turn, on the application of the circuit. For amplifiers with very small signal, the dc Q point could move clear to the edge of saturation and still function. On the other hand, a large-signal amplifier might tolerate only a 20 percent change in the Q point before clipping occurs.

For our purposes, we shall consider that the transistor can move its operating point just to the edge of saturation before failing. This will allow a simple approach to the problem. Thus the first information to be derived is that of the static, or quiescent, operating condition of the circuit.

Once the dc voltages and currents are known for ambient-temperature conditions (25°C), we can easily calculate how much the transistor currents must increase to drive it to saturation. When this information is known, we can apply some simple rules to determine how much the temperature must rise to cause such a condition.

An example will serve to show the method and procedure. Figure 11-13a shows a circuit using a Motorola 2N2043A transistor; its temperature characteristics are given in Fig. 11-13b. A dc analysis of this

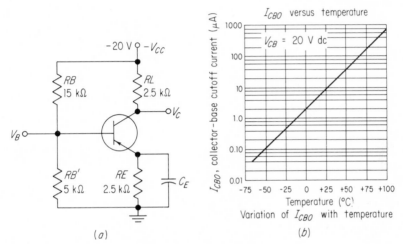

(a)

(b)

FIG. 11-13. (a) Temperature-calculation example; (b) I_{cbo} versus temperature. (By permission of Motorola Semiconductors, Inc.)

circuit shows that V_{CE} is 10 V, indicating that the transistor is biased at the center of the dc load line. Collector current is 2 mA; so 5 V is dropped across both RE and RL. Thus V_C, measured to ground, is quiescently −15 V.

If the transistor goes just to the edge of saturation, its internal resistance becomes zero, and V_C is −10 V, again referred to ground. The change in voltage across RL, divided by RL, will give the change in I_C necessary to cause the voltage change across RL.

$$\Delta I_C = \frac{\Delta E_{RL}}{RL} = \frac{5}{2.5 \text{ k}\Omega} = 2\text{-mA increase}$$

Thus collector current must double to drive the transistor to saturation.

The next step is to determine the path taken by I_{cbo} to determine if any part of it becomes amplified. If any significant part of it flows across the base-emitter region, it becomes amplified by the transistor, and $(\beta + 1) \times I_{cbo}$ will be caused to exist in the collector. The base-emitter equivalent circuit, shown in Fig. 11-14, will allow us to determine how much I_{cbo} flows through RB' and how much flows through the emitter resistance RE_{eq}. Assume $\beta_{dc}(h_{FE})$ is 100 for this transistor, which is the maximum given by the manufacturer for the 2N2043A.

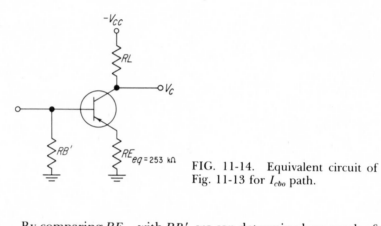

FIG. 11-14. Equivalent circuit of Fig. 11-13 for I_{cbo} path.

By comparing RE_{eq} with RB', we can determine how much of the leakage current flows in each leg, using the following relations:

$$RE_{eq} = (\beta + 1)RE \cong 253 \text{ k}\Omega$$

$$I_{cbo(RE_{eq})} = \frac{RB'}{RE_{eq} + RB'} = \frac{5 \text{ k}\Omega}{258 \text{ k}\Omega} = 0.0194 \cong 0.02, \text{ or } \tfrac{1}{50}$$

Hence about 50 times as much leakage current flows through RB' as through RE_{eq}. Only that part flowing across the emitter junction is amplified. This part, then, is the amount of the total leakage current that is beta times greater.

We now know that the total I_{cbo} must increase by 2 mA to cause the circuit to go to saturation because of a temperature increase. Using the following relation, we can determine the actual increase of I_{cbo} necessary to just cause this condition.

$$\Delta I_{cbo'} = (\beta + 1)\left(\Delta I_{cbo} \times \frac{RB'}{RB' + RE_{eq}}\right)$$

where $\Delta I_{cbo'}$ = total leakage increase, including amplified portion
ΔI_{cbo} = actual leakage increase due to temperature increase only

Solving for the unknown, ΔI_{cbo},

$$\Delta I_{cbo} = \frac{\Delta I_{cbo'}}{(\beta + 1)RB'/(RB' + RE_{eq})} = \frac{2 \text{ mA}}{101 \times 0.0194} \cong 1.02 \text{ mA}$$

From the temperature characteristics of Fig. 11-13, we see that I_{cbo} must increase from 9 μA at room temperature to 9 μA + 1.02 mA at the failure point. The temperature at which this occurs is read from the graph as about 110°C, where the running parameter intersects the 1-mA line (off scale). This is an increase of 85°C. Thus, if the temperature curves are available, we can use them to determine the temperature response for a given circuit rather easily.

If the curves are not available, a close approximation can be obtained by the following formula:

$$I_{cbo'} = I_{cbo}(2^y) \qquad \text{assuming a doubling of } I_{cbo} \text{ for every } 10°C \text{ rise}$$

where $y = \dfrac{\Delta T}{10}$

$\quad I_{cbo}$ = normal rated leakage current
$\quad I_{cbo'}$ = leakage current at upper temperature limit
$\quad \Delta T$ = increase of temperature above 25°C (77°F)

Solving for the exponential,

$$2^y = \frac{I_{cbo'}}{I_{cbo}} = \frac{2 \text{ mA}}{9 \mu\text{A}} = 222$$

This tells us that 2, raised to some power, equals 222.

$$y = \frac{\ln 222}{\ln 2} = 7.79 \cong 7.8$$

Since $y = \Delta T/10$, we can solve this relation for ΔT, which is the unknown.

$$\Delta T = y \times 10 = 7.8 \times 10 = 78°C$$

The temperature rise, then, is 78°C above ambient (25°C). The maximum temperature to just cause this condition (saturation) is the sum of 25 and 78, or 103°C.

Note that the two methods give slightly different results. This is due to the fact that the value of I_{cbo} does not exactly double for every

10°C rise in temperature. However, the approximation is quite close enough for most practical cases. Since this is a germanium transistor, its maximum operating temperature is about 85°C, and it would fail at this temperature before reaching the calculated figure. In other words, the circuit is very stable up to the maximum allowed.

If the amount of increase in I_{cbo} is known accurately, this figure can be substituted in the exponential of the above expression. As an example, the 2N2043A actually has an increase of I_{cbo} that doubles every 11.4°C.

$$2^y = \frac{I_{cbo'}}{I_{cbo}} = 222 \qquad \text{as before}$$

$$2^y = 222 \qquad \text{so } y = \frac{\ln 222}{\ln 2} = 7.8$$

$$y = \Delta T / 11.4$$

$$\Delta T = y \times 11.4 = 7.8 \times 11.4 = 88.9°C \text{ rise}$$

$$88.9 + 25 = 113.9°C$$

This agrees very well with the value derived from the curves, which was 110°C.

Let us now look into a circuit that is not at all temperature-stable. This should give a better feeling for the factors that cause temperature instability. The circuit in Fig. 11-15 is just such a circuit, and the

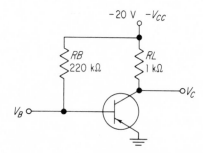

FIG. 11-15. Circuit example for poor thermal response.

transistor is the same as that used in the preceding example. With a beta of 100, the base current is 91 μA. From the curves shown in the preceding example, I_{cbo} is 9 μA at room temperature. So the total base current is the sum of 91 and 9, or 100 μA. We include I_{cbo} in this case because it is a reasonable part of the total base current.

Collector current is beta times the base current, or 10 mA. The drop across the load resistor is I_C times RL, or 10 V. Our transistor, then, is biased at the midpoint of the dc load line. If it were to go to saturation, I_C would be limited by RL only, and this would be 20 mA.

Now, I_C must increase by 10 mA to produce saturation. The question arises: By how much must I_{cbo} increase to cause an increase in collector current of 10 mA? In this circuit, I_{cbo} must flow from the collector to the base *and out the transistor by way of the emitter lead.* Thus not only does I_{cbo} itself flow in the collector, but an amplified amount, $\beta \times I_{cbo}$, must also flow in the collector. If we consider the normal collector current and the total effect of I_{cbo}, the total collector current is

$$I_C = \beta I_B + (\beta + 1)I_{cbo}$$

In our case, we know that βI_B is 10 mA. However, when the transistor is driven to saturation by an increase in the temperature, $(\beta + 1)I_{cbo}$ must also be 10 mA. The *increase* in total collector current, ΔI_C, is equal to $(\beta + 1)I_{cbo}$. So in our case

$$I_{cbo} = \frac{I_C}{\beta + 1} = \frac{10 \text{ mA}}{101} = 99 \text{ } \mu A$$

This tells us that if I_{cbo} increases by 99 μA, the transistor will just be driven to saturation. Thus, if base current increases from 100 μA total at 25°C to 199 μA at the elevated temperature, the transistor has just reached saturation.

From the curve for this transistor, in Fig. 11-13, we read up from 9 μA to $9 + 99 = 108$ μA, and this occurs at about 65°C. The allowed temperature rise is from 25 to 65, a 40° change.

By using our formula, we can arrive at essentially the same information if the curves are not available. Assume that I_{cbo} doubles for every 11.4°C rise.

$$2^y = I_{cbo'}/I_{cbo} \qquad \text{where } y = \Delta T/11.4$$

$$2^y = 99/9 = 11$$

$$2^y = 11 \qquad \text{and} \qquad y = \frac{\ln 11}{\ln 2}$$

$$y = 3.459$$

$$y = \Delta T/11.4$$

so $\qquad \Delta T = y \times 11.4 = 3.459 \times 11.4 = 39.4°C \text{ rise}$

Again, this agrees well with the graphical determination. The maximum temperature for this circuit, then, is $25 + 39.4 = 64.4°C$. Note that this circuit has poor temperature characteristics. One reason is that I_{cbo} is amplified by the transistor. Another reason is that the resistors are relatively large. If RL, for instance, were a 5-kΩ resistor and the other were adjusted accordingly, the circuit would then fail with only a 15°C rise. Thus many factors affect the temperature stability of a transistor circuit.

11-9 SILICON TEMPERATURE RESPONSE

The foregoing discussion has been concerned with the temperature characteristics of germanium transistors, for the most part. It should be noted that a silicon transistor, while quite similar to a germanium transistor in its response to a temperature change, reacts in a somewhat different manner. For silicon, the leakage current I_{cbo} increases at a rate that doubles for about every 6°C rise in temperature. Also, in the usual case, a silicon unit has far less leakage current than a typical germanium transistor at room temperature.

For example, one germanium transistor (alloy-junction type) has a specified I_{cbo} of 3 μA at 25°C. Compare this with the following silicon units: a 2N2714 (NPN silicon) rated at 0.5 μA and a 2N2693 (NPN silicon) rated at 10 *nano*amperes (0.01 μA) at 25°C. In the latter case, 10 nA can almost be considered an insignificant amount of leakage current.

One important result of using transistors with very low values of leakage current is that the biasing resistors may be very much larger in value, thus reducing the current drain on the power supply. Even the load resistor can be made much larger, with values in the vicinity of 50 kΩ not uncommon.

Except for these differences, a circuit using a silicon transistor is treated in the same manner as one using a germanium unit. With suitable modification, the discussion in this chapter applies to both types.

QUESTIONS AND PROBLEMS

11-1 State the two purposes of a biasing network.

11-2 Discuss the general temperature response of each of the three basic circuit configurations.

11-3 A given germanium transistor has a leakage current of 6 μA at 25°C. When the temperature is increased to 85°C, what is the new value of leakage current, approximately? Select the correct answer.

(a) 3 μA   (b) 6 μA
(c) 48 μA   (d) 192 μA
(e) 384 μA   (f) 768 μA

11-4 The circuit shown is a linear amplifier. The germanium transistor and the circuit have the characteristics given. Normal collector current is 0.6 mA at 25°C. The transistor is to be operated at 65°C. At this temperature what is the voltage from collector to emitter? Select the correct answer.

(a) 7.7 V   (b) 4.3 V
(c) 6.0 V   (d) 0.0 V
(e) 3.6 V   (f) 9.1 V

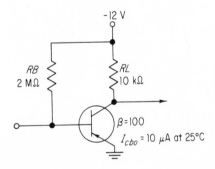

11-5 A 200-mW transistor is operated in the circuit shown. It is to be operated at 65°C. The manufacturer gives the derating information as being 3.2 mW/°C. Will it perform satisfactorily?

(a) Yes   (b) No

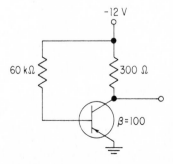

11-6 What is the S factor of the circuit shown? Select the correct answer.

(a) 1   (b) 83.5
(c) 16.5   (d) 100

Ch 11 Q4

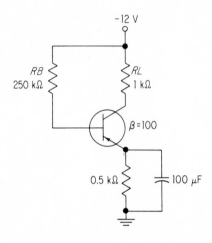

11-7 In the circuit shown, what is P_C? Select the correct answer.

(a) 200 mW   (b) 0.1 W
(c) 48 mW   (d) 24 mW

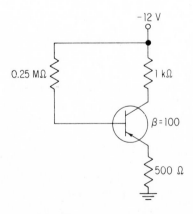

11-8   The transistor in the circuit shown is rated at 200 mW. For the conditions given, is it operating at a safe level of dissipation?

(a) Yes   (b) No

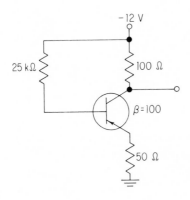

11-9   Select the correct answer. Silicon transistors have better temperature response because
(a) their I_{cbo} increases at a faster rate
(b) their leakage current is much lower to start with (25°C)
(c) they operate at lower temperatures
(d) they have a natural built-in heat sink

11-10   Select the correct answer. Leakage currents in typical germanium transistors increase at the approximate rate of
(a) 2 times for every 10°F
(b) 2 times for every 10°C
(c) 10 times for every 2°C
(d) ⁹⁄₅ times for every 32°F

11-11   Select the correct answer. The detrimental effect of increasing I_{cbo} is that
(a) the frequency response becomes worse
(b) the collector dissipation becomes less
(c) the ac load line shifts more perpendicular
(d) the dc operating point moves up the load line

11-12 The circuit shown is used to measure leakage current at 25°C. The transistor being measured has a rated I_{cbo} of 3 μA. The meter measures about 300 μA. Explain in detail the discrepancy in values.

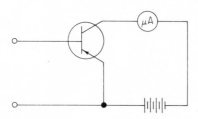

11-13 The circuit shown is used to measure leakage current. The meter reads 500 μA. What is the value of I_{cbo}? Given that $V_{CC} = 10$ V, $\alpha = 0.992$, $T_A = 25$°C, Q_1 is germanium.

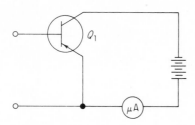

11-14 Select the correct answer. Leakage current in a typical silicon transistor increases at the approximate rate of
(a) 2 times for every 10°F
(b) 2 times for every 10°C
(c) 2 times for every 6°C
(d) ⅝ times for every 10°C

11-15 A germanium transistor has a rated leakage current of 5 μA at 25°C. Find the value of I_{cbo} at 45°C.

11-16 Refer to Question 11-15. Find the value of I_{cbo} at 85°C.

11-17   Refer to Question 11-15. Find the value of I_{cbo} at 50°C (approximate).

11-18   A silicon transistor has a rated leakage current of 2 μA at 25°C. Find the value of I_{cbo} at 49°C.

11-19   Refer to Question 11-18. Find the value of I_{cbo} at 67°C.

11-20   Refer to Question 11-18. Find the value of I_{cbo} at 51°C.

11-21   A silicon transistor has a listed I_{cbo} at 25°C of 100 nA. Find the value of I_{cbo} at 91°C.

11-22   A silicon transistor has a listed I_{cbo} at 25°C of 10 nA. Find the value of I_{cbo} at 61°C.

11-23   Refer to Fig. 11-13. Determine the quiescent power dissipation P_C.

11-24   Refer to Question 11-5. A 200-mW transistor is to be used that has a 1.5 mW/°C derating factor. Will the circuit now perform satisfactorily?

CHAPTER 12
ANALYSIS OF LINEAR CIRCUITS

Up to this point, the individual steps in an analysis of basic circuits have been presented. The three conditions of the circuit that must be known in order to describe the total circuit action are the dc state of the circuit, the large-signal response, and the small-signal response. Because these circuit conditions have been treated more or less separately, it would be advantageous to provide complete analyses of several circuit examples, and that is the purpose of this chapter.

Although graphical analysis could be applied here, it is often difficult to obtain the characteristic curves for a particular transistor, and this difficulty sometimes precludes this method. Besides, a presentable analysis can be performed without using characteristic curves, and so here we shall use a nongraphical approach.

12-1 FIXED-BIAS CIRCUIT

Our first example is used for its simplicity, and shows the necessary methods in complete form, without becoming too complex. For the present purpose, an ideal analysis system will allow one to analyze a circuit with only the information shown on a typical schematic diagram, plus some knowledge of the transistor characteristics. Figure 12-1 shows only this information for the fixed-bias circuit. In this

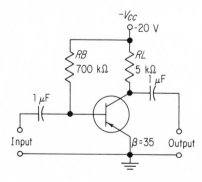

FIG. 12-1. Analysis example of fixed-bias circuit.

chapter we shall use the ERCA equations for our dc analysis exclusively, since it is the one simple system that will work for all circuits.

Our first step, then, is to determine the quiescent condition of the circuit. We recall that the ERCA relation for this circuit is that RL plus R_{CB} must be smaller than RB by the factor beta. The proper equation is

$$R_{CB} = \frac{RB}{\beta} - RL = \frac{700 \text{ k}\Omega}{35} - 5 \text{ k}\Omega = 15 \text{ k}\Omega$$

The simplified dc equivalent circuit can now be drawn and solved for the voltages and currents that exist quiescently.

$$I_C = \frac{V_{CC}}{R_t} = \frac{20}{20 \text{ k}\Omega} = 1 \text{ mA}$$

$$-V_C = \frac{R_{CB}}{RL + R_{CB}} \times -V_{CC} = \frac{15 \text{ k}\Omega}{20 \text{ k}\Omega} \times -20 = -15 \text{ V}$$

Base current can be determined by either of two equations.

$$I_B = \frac{V_{CC}}{RB} = \frac{20}{700 \text{ k}\Omega} = 0.02857 \text{ mA} \cong 29 \ \mu\text{A}$$

$$I_B = \frac{I_C}{\beta} = \frac{1 \text{ mA}}{35} = 0.02857 \cong 29 \ \mu\text{A}$$

By calculating gamma we can easily visualize the point along the dc load line at which the transistor is biased.

$$\gamma = \frac{R_{CB}}{RL + R_{CB}} = \frac{15 \text{ k}\Omega}{20 \text{ k}\Omega} = 0.75 \quad \text{or} \quad \frac{V_{CE}}{V_{CC}} = \frac{15 \text{ V}}{20 \text{ V}} = 0.75$$

Thus the transistor is biased one-quarter of the way up the load line from cutoff. Since this example is very simple, the analysis is just as simple, and the dc operating condition of the circuit has been completely specified.

The large-signal response of this circuit is quite straightforward because of its simplicity. If the transistor is driven to saturation, the collector voltage falls to ground, nearly. If the transistor is driven to cutoff, the collector voltage is equal to V_{CC}, in this case -20 V. Thus the maximum possible collector swing is from ground to -20 V.

However, since the quiescent value is -15 V, the maximum collector swing will not be of symmetrical shape. The maximum possible symmetrical collector swing is 2 times the excursion *to the nearest limit.* The quiescent collector voltage is -15 V, and the nearest limit is cutoff. This is a 5-V swing; so the maximum possible symmetrical excursion is 2×5, or 10 V, peak to peak. This is typical of circuits with a gamma of something other than 0.5. For maximum large-signal collector swing, the transistor should be biased in the center of the ac load line. If, however, the transistor is never to be driven this hard, it is often advantageous to bias it nearer to the cutoff point, to conserve power drain. This is usually the case with preamplifiers and other small-signal amplifiers.

We shall now consider the small-signal response of the circuit. The factors that must be determined are the voltage gain, the input resistance, and the approximate frequency response. Since the emitter is tied to ground, we must first determine the value of r_e. Recalling that this relationship is $26/I_E$ (mA), we can easily determine its value.

$$r_e = \frac{26}{I_E} = \frac{26}{1} = 26 \ \Omega$$

This is the effective ac resistance of the emitter. Knowing the value of r_e, we can now determine the approximate voltage gain.

$$A_v = \frac{RL}{r_e} = \frac{5 \text{ k}\Omega}{26} = 192$$

The voltage gain, then, is about 192, and if a 1-mV signal were impressed upon the input, the output would be a 192-mV excursion.

The input resistance of this circuit is very low, since there is no external resistance in the emitter lead. By momentarily neglecting RB, we can quickly determine R_{BE}.

$$R_{BE} = (\beta + 1)r_e = (35 + 1) \times 26 = 936 \ \Omega$$

The total input resistance r_i must include all shunt paths, and RB must be included in this calculation for good accuracy if RB is similar in value to R_{BE}. Since 700 kΩ is very much larger than 936 Ω, we can ignore RB and make almost no error. In this case, the error is less than 0.2 percent.

We have left only the frequency determination for this circuit. If the manufacturer of the transistor has given the alpha cutoff frequency as 700 kHz, we must determine the beta cutoff frequency from this.

$$F_\beta = \frac{F_\alpha}{\beta} = \frac{700 \text{ kHz}}{35} = 20 \text{ kHz}$$

The highest frequency possible for this transistor in the common-emitter configuration is therefore 20 kHz, and it is probably operating at a lower frequency because of stray capacity in the circuit.

The low-frequency response of this circuit is limited by the coupling capacitors. Since we do not know the load to which the output coupling capacitor is connected, we shall ignore its effect on the frequency response. However, the input capacitor will certainly cause a reduction in gain on the low side of the response curve. The point that we are concerned with is the frequency where X_C equals the r_i of the transistor. Assume that r_i is equal to 1000 Ω, and C is a 1-μF (microfarad) capacitor.

$$F_{\text{low}} = \frac{1}{2\pi RC} = \frac{1}{6.28 \times 1 \text{ k}\Omega \times 1 \text{ }\mu\text{F}} \cong 160 \text{ Hz}$$

Note that the low-frequency limit for the circuit is relatively high. One reason for this is that the input resistance of the transistor is very low. If the coupling capacitor were made larger, the low-frequency response would improve. Or, making r_i greater in value would do the same thing.

Let us now summarize what we have determined about this circuit. The collector voltage with no signal applied is −15 V, and the quiescent collector current is 1 mA. The circuit gamma is 0.75; so the transistor is biased toward cutoff, one-quarter of the way up the load line between cutoff and saturation. The maximum possible symmetrical collector excursion is 10 V, peak to peak. The input resistance to a signal is on the order of 1000 Ω, and the voltage gain is 192. The frequency response of the circuit as shown is from about 160 Hz to about 20 kHz. Thus we have described all the measurable quantities of this circuit, as we set out to do, and with a minimum of effort.

12-2 UNIVERSAL CIRCUIT

The second example is shown in Fig. 12-2. This is a typical universal circuit, with an additional emitter resistor used to increase the input resistance. To determine the quiescent conditions in this circuit, we shall first derive the ERCA equation. RB is greater than $RL + R_{CB}$ by the same amount that RB' is greater than RE_{total}. Solving this relationship for R_{CB} will yield the ERCA equation for this circuit.

$$R_{CB} = \frac{RB \times RE}{RB'} - RL = \frac{15 \text{ k}\Omega \times 1 \text{ k}\Omega}{5 \text{ k}\Omega} - 1 \text{ k}\Omega$$

$$= 3 \text{ k}\Omega - 1 \text{ k}\Omega = 2 \text{ k}\Omega$$

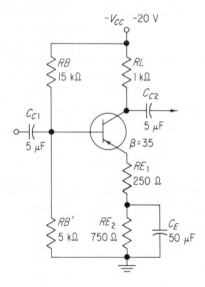

FIG. 12-2. Analysis example of universal circuit.

Now I_C, V_{CE}, V_C, and V_E can be easily calculated. The collector current is, of course, determined by the total voltage across the total series resistance.

$$-I_C = \frac{-V_{CC}}{RL + R_{CB} + RE_1 + RE_2} = \frac{-20}{4.0 \text{ k}\Omega} = -5 \text{ mA}$$

The voltage drop across any of the resistors in the collector circuit or emitter circuit can now be found.

$$-V_E = -I_E \times RE_{\text{total}} = -5 \text{ mA} \times 1 \text{ k}\Omega = -5 \text{ V}$$

(Since V_{CC} is minus, V_E and V_C are minus.)

$$-V_C = -I_E \times (RE + R_{CB}) = -5 \text{ mA} \times (1 \text{ k}\Omega + 2 \text{ k}\Omega)$$

$$= -5 \text{ mA} \times 3 \text{ k}\Omega = -15 \text{ V}$$

$$V_{CE} = -V_C + V_E = -15 + 5 = -10 \text{ V}$$

$$E_{RL} = I_C \times RL = 5 \text{ mA} \times 1 \text{ k}\Omega = 5 \text{ V}$$

One final step concludes the quiescent analysis. It is always necessary to know gamma, for the bias of the transistor determines many other characteristics.

$$\gamma = \frac{V_{CE}}{V_{CC}} = \frac{10}{20} = 0.5$$

The transistor in this circuit is biased at the midpoint of the dc load line.

We can now proceed to find the large-signal and the small-signal responses of this circuit. As always, the two limits of operation are specified by the two extremes of transistor operation: cutoff and saturation. In this case, one must be careful to specify whether the dc or ac (signal) limits are called for.

The main concern is with the ac limits of operation, and so it is of great interest to determine how the emitter-bypass capacitor influences the circuit. For signal conditions the bypass capacitor represents a short circuit to variations of emitter current. Or, saying the same thing a different way, it holds the upper end of RE_2 at a constant dc voltage level for fast signal excursions.

An ac equivalent circuit of Fig. 12-2 would show that the upper end of RE_2 is at a dc level that might be considered to be provided by a separate power supply $V_{E'}$. The value of this voltage is

$$-V_{E'} = I_E \times RE_2 = -5 \text{ mA} \times 0.75 \text{ k}\Omega = -3.75 \text{ V}$$

This is an important idea, and it tells us that if the transistor is cut off momentarily, the most positive the emitter can go is -3.75 V, *not* ground. The voltage from collector to emitter, V_{CE}, is therefore somewhat less than the full range of V_{CC} when the transistor is cut off. Thus the ac swing of the collector is smaller than would be the case if the change were made very slowly.

$$V_{CE(max)} = -V_{CC} + V_{E'} = -20 + 3.75 = -16.25 \text{ V}$$

In the other direction, with the transistor driven toward saturation very rapidly, CE again influences the operation. Instead of V_C falling to ground, as it would in a circuit similar to that shown in Fig. 12-1, it can only fall to a voltage that is more negative than ground. With the transistor in saturation, and therefore R_{CB} nearly equal to zero, the voltage divider consisting of RL and RE_1 determines the value to which the collector will fall.

$$E_{RE'} = \frac{RE'}{RL + RE'} \times V_{CE(max)} = \frac{250}{1250} \times -16.25 = -3.25 \text{ V}$$

$$V_{E(max)} = (-3.25) + (-3.75) = -7 \text{ V}$$

In other words, the most positive that the collector can go is a value formed by the drop across RE_1 returned to -3.75 V, or -7.0 V. The maximum peak-to-peak collector excursion, then, is from -20 to -7 V, a 13-V swing. If either RE_1 or RE_2 were larger in value, the maximum collector swing would be even more restricted.

At this point, the small-signal response will be determined for this circuit. It is usually best first to determine r_e at the quiescent condition.

$$r_e = \frac{0.026}{I_E} = \frac{0.026}{5 \text{ mA}} = 5.2 \text{ } \Omega$$

The voltage gain A_v can be calculated next.

$$A_v = \frac{RL}{RE' + r_e} = \frac{1 \text{ k}\Omega}{255} = 3.9$$

The input resistance r_i is next, and this is divided into two determinations. The first of these is the parallel resistance of RB and RB', and the second is R_{ib}. The parallel resistance of RB and RB' we shall call RP.

$$RP = \frac{RB \times RB'}{RB + RB'} = \frac{5 \text{ k}\Omega \times 15 \text{ k}\Omega}{5 \text{ k}\Omega + 15 \text{ k}\Omega} = 3.75 \text{ k}\Omega$$

$$R_{ib} = (\beta + 1)(r_e + RE_1) = 36 \times 255.2 = 9.2 \text{ k}\Omega$$

The total input resistance r_i is the parallel value of these two figures.

$$r_i = \frac{RP \times R_{ib}}{RP + R_{ib}} = \frac{3.75 \text{ k}\Omega \times 9.2 \text{ k}\Omega}{3.75 \text{ k}\Omega + 9.2 \text{ k}\Omega} = 2.66 \text{ k}\Omega$$

Thus the total input resistance seen by the outside source is roughly 2.7 kΩ.

If the manufacturer of the transistor gave us the high-frequency information as $f_\alpha = 5$ MHz, we should have to convert f_α to f_β.

$$f_\beta = \frac{f_\alpha}{\beta} = \frac{5 \text{ MHz}}{35} = 143 \text{ kHz}$$

Because of the emitter feedback caused by the unbypassed emitter resistor, the upper frequency limit of the circuit is much higher than indicated by f_β. The following relationship yields an approximation of this limit for this kind of circuit.

$$f_2 \cong f_\beta + \frac{f_\alpha}{A_v}$$

where f_2 = upper limit of the circuit
f_β = upper limit of a simple CE circuit
f_α = upper limit of a CB circuit
A_v = voltage gain

In the case of the present circuit, f_2 is

$$f_2 \cong f_\beta + \frac{f_\alpha}{A_v} = 142 \text{ kHz} + \frac{5 \text{ MHz}}{3.9}$$

$$= 142 \text{ kHz} + 1280 \text{ kHz} \cong 1.42 \text{ MHz}$$

Hence, because of the reduced gain of this circuit, the frequency response is greatly increased.

The low-frequency response can be approximated in two steps. The way in which the reactance of the coupling capacitor compares with r_i is one determining factor. As before, the frequency where $X_C = r_i$ is the point at which gain begins to become appreciably less than at midband.

$$f = \frac{1}{2\pi X_C C_{C1}} = \frac{1}{6.28 \times 2.7 \text{ k}\Omega \times 0.000005} = 11.8 \text{ Hz}$$

Now, the frequency at which the reactance of CE becomes equal to RE_2 will also affect the low-frequency response.

$$f = \frac{1}{2\pi X_c CE} = \frac{1}{6.28 \times 750 \times 0.00005} = 4.24 \text{ Hz}$$

Neglecting the effect of C_{C2} on the following stage, the true low-frequency response is very nearly equal to the larger of the two values, or about 12 Hz.

12-3 DIFFERENTIAL AMPLIFIER

The example in Fig. 12-3 is often called a *differential amplifier.* As shown, it would ordinarily require a rather elaborate analysis to describe both the dc and ac conditions. The circuit appears to be somewhat more complex than others we have investigated, but it is really quite simple, as we shall see. It performs several functions, converting a single-ended signal to a push-pull signal, thus acting as its own phase inverter. Although it is not immediately apparent, all three

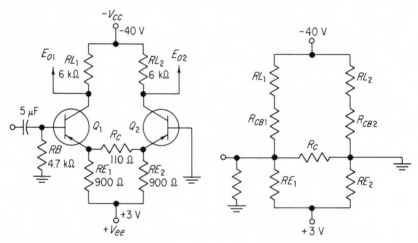

FIG. 12-3. Analysis example of differential circuit.

FIG. 12-4. Dc equivalent circuit differential amplifier.

basic configurations are found in this circuit: common collector, common base, and common emitter. It will not amplify "common-mode" signals from ripple on the power supply; thus it is said to have "common-mode rejection."

To illustrate the circuit operation, we can begin by analyzing it for its quiescent conditions. The dc equivalent circuit is shown in Fig. 12-4. Considering one transistor at a time, we can use our simplified method to determine the dc conditions. Ignoring the 110-Ω resistor for the moment, consider first that the base voltage must be very nearly 0 V with respect to ground. Since the current flowing through

RB, the 4.7-kΩ base resistor, is very small, the drop across it must be accordingly small. Since this is true, the emitter must also be very nearly 0 V, too.

This leads us to the ERCA relation for this circuit. The load resistor of Q_1 plus R_{CB} is related to $-V_{CC}$ by the same ratio that RE is related to V_{EE}. Solving this for R_{CB} gives us

$$R_{CB} = \frac{V_{CC} \times RE}{V_{EE}} - RL = \frac{40 \times 0.9 \text{ k}\Omega}{3} - 6 \text{ k}\Omega$$

$$= 12 \text{ k}\Omega - 6 \text{ k}\Omega = 6 \text{ k}\Omega$$

The dc condition of the circuit of Q_1 can now be determined.

$$V_C = \frac{R_{CB}}{RL + R_{CB}} \times V_{CC} = \frac{6 \text{ k}\Omega}{6 \text{ k}\Omega + 6 \text{ k}\Omega} \times (-40) = -20 \text{ V}$$

Collector current is simply V_{CE} divided by R_{CB}. ($V_{CE} = V_C$, since the emitter is at ground.)

$$I_C = \frac{V_{CE}}{R_{CB}} = \frac{20}{6 \text{ k}\Omega} = 3.33 \text{ mA}$$

Finding gamma will complete the dc analysis of this circuit.

$$\gamma = \frac{R_{CB}}{RL + R_{CB}} = \frac{6 \text{ k}\Omega}{12 \text{ k}\Omega} = 0.5$$

Now, since Q_2 is in nearly the same circuit, it can be assumed that its quiescent values are the same as those of Q_1. Thus V_C (and V_{CE}) are −20 V, I_C is 3.33 mA, and gamma is 0.5 for both transistors.

At this point, we must decide whether the 110-Ω coupling resistor R_c will disturb the previous calculations. If the bases of Q_1 and Q_2 are essentially at ground, the emitters are at, perhaps, +0.1 V, because of the small drop across the emitter-base junction. The resistor R_c is across a difference of potential of 0 V since its left side is at 0.1 V, and its right side is at 0.1 V. Thus no current flows through it quiescently, and it in no way upsets the previous calculations.

It will now be helpful to describe how the circuit works when a signal is applied to the input, instead of performing a straight analysis. Assume that the input is a small step voltage, rather than a sinusoid. This will allow small voltage levels to be considered, rather than a continuously variable voltage. If the input goes from 0 to −0.1 V, the input transistor is turned on harder. Since the emitter is not clamped to any particular voltage, it will follow this change. The base going from ground to −0.1 V will cause the emitter to go from +0.1 to 0 V.

We must now consider the coupling resistor R_c, for there is 0.1 V across it. With 0.1 V across it, the current through it will be determined by the voltage and the value of the resistor.

$$I = \frac{E_{RC}}{R_c} = \frac{0.1}{110} = 0.91 \text{ mA}$$

Since the left end of R_c is the more negative, current will flow from left to right. This is shown in Fig. 12-5. The origin of this current is the Q_1 collector circuit, since Q_1 is turned on harder by the input signal. Most of this 0.91 mA will flow down through RE_2, and

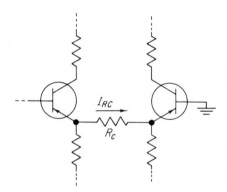

FIG. 12-5.   Current through R_c.

a very small part of it will try to oppose the normal base current of Q_2. This tends to reduce the current in Q_2, for as the base current is reduced, the collector current must also be reduced.

Now Q_1 collector current has increased, and Q_2 collector current has decreased, and it is interesting to determine the amount of these changes. Q_1 collector current increases by 0.91 mA, and this *increase* in Q_1 current is the same as the amount of *decrease* in Q_2 current. This can be proved as shown in Fig. 12-6.

With zero input current, 3.33 mA flows down through the 900-Ω resistor from the source Q_2. The upper end of the 900-Ω resistor, RE_2, is clamped to ground by the base of the transistor. When 0.91 mA is sent into the network from Q_1, the voltage across

RE_2 remains constant, and thus the current through it remains constant. Since 0.91 mA is delivered by Q_1, then $I_C Q_2$ must decrease, so that Ohm's law can be satisfied, and the drop across RE_2 remains constant. Hence, as $I_C Q_1$ increases, $I_C Q_2$ decreases, and by the same amount.

Quiescently, Q_1 collector voltage, measured to ground, is -20 V. Because of the input signal, it must now be less negative.

$$-V_{C'} = -V_{CC} + E_{RL} = -V_{CC} + (I_{C'} RL) = -40 + 25.44 = -14.56 \text{ V}$$

The change in collector voltage, then, is from -20 to about -14.6 V, a change of 5.4 V. Knowing this, we can determine the voltage gain by the use of the input-output relationship.

$$A_v = \frac{e_{\text{out}}}{e_{\text{in}}} = \frac{5.44}{0.1} = 54.4$$

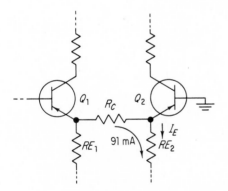

FIG. 12-6. I_{Rc} flows through RE_2.

An alternative way to determine the gain is to use the RL/RE relationship. In this circuit, R_c is the effective emitter resistance.

$$A_v = \frac{RL}{R_c} = \frac{6 \text{ k}\Omega}{110} = 54.5$$

This proves that both methods of determining the gain agree very well.

If the collector of Q_1 falls toward ground by 5.44 V, the collector of Q_2 must rise toward $-V_{CC}$ by 5.44 V. Thus the difference between output 1 and output 2 is 10.8 V. The collector-to-collector voltage gain is therefore 2×54.4, or 108. In other words, the input voltage swing, from one extreme to the other, is 0.1 V; the output voltage swing, collector to collector, is 10.8 V, an increase of 108.

The limits of operation of, for instance, the collector of Q_1 can easily be determined. If the transistor is cut off, the collector voltage, referred to ground, is equal to $-V_{CC}$, or -40 V. This is one of the limits. The other limit is reached when the transistor is just at saturation. By making a simplifying assumption we can easily determine the limit at this point of operation. Since the emitter of Q_2 is for all practical purposes clamped to ground, we can consider only the coupling resistor as the emitter resistor of Q_1. Thus RL and R_c form a voltage divider that will set the level of V_C as the transistor is just driven to saturation.

$$V_{c'} = \frac{R_c}{RL + R_c} \times (-V_{CC}) = \frac{0.11 \text{ k}\Omega}{6 \text{ k}\Omega + 0.11 \text{ k}\Omega} \times (-40) = -0.72 \text{ V}$$

Thus the collector can range between -40 and -0.72 V, a nearly 40-V excursion, as its maximum range.

Let us now approximate some of the small-signal parameters of this circuit. We have already determined the voltage gain, but the total input resistance r_i is not known at present. The overall input resistance is the shunt value of RB and the R_{ib} of the transistor, viewed from the base to signal ground.

If the voltage at the emitter of Q_1 changes slightly because of a signal, the voltage across RE_1 and across R_c must change. The total resistance seen by the signal at the emitter is the parallel resistance of these two resistors. If the minimum published value of beta is 35, we can use this to determine the R_{ib}.

$$R_{ib} = (\beta + 1) \frac{RE_1 \times R_c}{RE_1 + R_c} = 36 \times \frac{900 \times 110}{900 + 110} = 36 \times 98 \cong 3.5 \text{ k}\Omega$$

The total input resistance is the shunt combination of R_{ib} and RB.

$$r_i = \frac{RB \times R_{ib}}{RB + R_{ib}} = \frac{4.7 \text{ k}\Omega \times 3.5 \text{ k}\Omega}{4.7 \text{ k}\Omega + 3.5 \text{ k}\Omega} = 2 \text{ k}\Omega$$

The input signal source, then, sees a total input resistance of about 2 kΩ.

The high-frequency response of this circuit depends primarily upon the transistor characteristics and the construction methods used.

The low-frequency response, however, is determined by the coupling capacitor and r_i.

$$f = \frac{1}{2\pi X_C C} = \frac{1}{6.28 \times 2 \text{ k}\Omega \times 5 \text{ }\mu\text{F}} = 15.9 \text{ Hz}$$

Since this is the only reactance in the circuit, it is the only limiting factor, and the low-frequency response just begins to fall off at 16 Hz.

12-4 CASCADED AMPLIFIER

A final example is shown in Fig. 12-7a, and it can be seen to be a two-stage amplifier. Both transistors are operating in the universal circuit. Each has a 100-Ω resistor in the emitter circuit to improve the linearity. The output amplitude is variable because of the 10-kΩ potentiometer. The circuit has an overall voltage gain of about 600, with the low-frequency response extending to about 30 Hz.

To determine the overall response of this circuit, we must first determine the quiescent operating condition. The most straightforward way of doing this is to use the ERCA equation for the universal circuit. For the dc condition of the circuit we shall use the combined value of RE and RE'. This means that RE will be treated as a single 1.1 kΩ resistor. The ERCA equation for this universal circuit is derived from the fact that RB is greater than RL plus R_{CB} by the same amount that RB' is greater than RE_{total}. Solving this for the value of R_{CB} will allow us to determine the operating condition of the input circuit, or stage 1.

$$R_{CB} = \frac{RB \times RE}{RB'} - RL = \frac{160 \text{ k}\Omega \times 1.1 \text{ k}\Omega}{10 \text{ k}\Omega} - 8.2 \text{ k}\Omega$$

$$= 17.6 \text{ k}\Omega - 8.2 \text{ k}\Omega = 9.4 \text{ k}\Omega$$

With a value determined for R_{CB}, the gamma can be quickly determined for this circuit.

$$\gamma = \frac{R_{CB}}{RL + R_{CB} + RE} = \frac{9.4 \text{ k}\Omega}{8.2 \text{ k}\Omega + 9.4 \text{ k}\Omega + 1.1 \text{ k}\Omega}$$

$$= \frac{9.4 \text{ k}\Omega}{18.7 \text{ k}\Omega} = 0.5$$

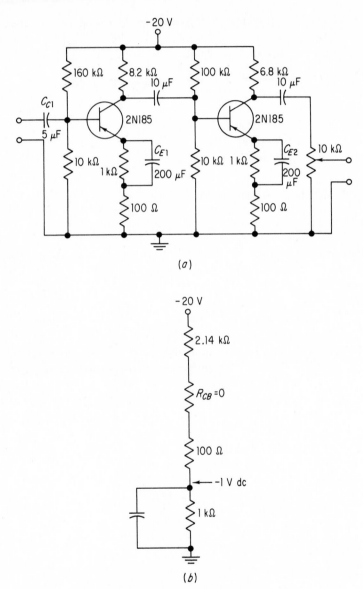

FIG. 12-7. Two-stage amplifier.

With a gamma of 0.5, the input transistor is biased at the center of the dc load line. The dc voltages at the collector, base, and emitter can now be calculated. First, it is convenient to determine the collector current.

$$-I_C = \frac{-V_{CC}}{R_{total}} = \frac{-20}{8.2 \text{ k}\Omega + 9.4 \text{ k}\Omega + 1.1 \text{ k}\Omega}$$

$$= \frac{-20}{18.7 \text{ k}\Omega} = -1.07 \text{ mA}$$

The collector voltage V_C can be arrived at in a number of ways. One of these is to use the collector current and the total resistance from collector to ground.

$$-V_C = -I_C \times (R_{CB} + RE) = 1.07 \text{ mA} \times (9.4 \text{ k}\Omega + 1.1 \text{ k}\Omega) = -11.2 \text{ V}$$

If $I_E \cong I_C$, then

$$-V_E = -I_E \times RE = -1.07 \text{ mA} \times 1.1 \text{ k}\Omega = -1.18 \text{ V}$$

If the drop from base to emitter is on the order of 0.1 V, the base voltage is, for all practical purposes, the same as the emitter voltage. The operating condition of the input amplifier has been easily solved for, and it is convenient to solve next for the dc condition of the output stage. This, of course, is accomplished in the same manner as for the first stage.

$$R_{CB} = \frac{RB \times RE}{RB'} - RL = \frac{100 \text{ k}\Omega \times 1.1 \text{ k}\Omega}{10 \text{ k}\Omega} - 6.8 \text{ k}\Omega$$

$$= 11 \text{ k}\Omega - 6.8 \text{ k}\Omega = 4.2 \text{ k}\Omega$$

$$\gamma = \frac{R_{CB}}{RL + R_{CB} + RE} = \frac{4.2 \text{ k}\Omega}{6.8 \text{ k}\Omega + 4.2 \text{ k}\Omega + 1.1 \text{ k}\Omega}$$

$$= \frac{4.2 \text{ k}\Omega}{12.1 \text{ k}\Omega} = 0.35$$

The output transistor is not biased at the center of the dc load line, but rather it is biased about two-thirds of the way up from cutoff. Because the input stage has amplified the signal, it is desirable to have the transistor biased at the center of the ac load line. With large values of RE, a gamma of 0.35 is very nearly the center of the ac load line.

Now the dc voltages and currents can easily be determined.

$$-I_C = \frac{-V_{CC}}{R_{total}} = \frac{-20}{6.8 \text{ k}\Omega + 4.2 \text{ k}\Omega + 1.1 \text{ k}\Omega}$$

$$= \frac{-20}{12.1 \text{ k}\Omega} = -1.65 \text{ mA}$$

$$-V_C = -I_C \times (R_{CB} + RE) = -1.65 \text{ mA} \times 5.3 \text{ k}\Omega = -8.76 \text{ V}$$

$$-V_E = -I_E \times RE = -1.65 \text{ mA} \times 1.1 \text{ k}\Omega = -1.82 \text{ V} \qquad I_E \cong I_C$$

The signal response of the circuit can now be determined. However, the large- and small-signal responses are interrelated, and cannot be easily separated when two or more circuits are cascaded. Hence we must modify our basic method slightly.

It is first necessary to find the total input resistance of each stage, r_i. Only then can we determine the large-signal response. Taking the input transistor first, the total input resistance is the input resistance of the transistor itself, R_{ib}, in parallel with the parallel resistance of RB and RB', RP. For this example, we shall assume that the transistor beta is 35, with $I_E \cong 1$ mA and $r_e \cong 26$ Ω.

$$R_{ib} = (\beta + 1)(r_e + RE_{\text{unbypassed}}) = 36 \times 126 = 4.5 \text{ k}\Omega$$

$$RP = \frac{RB \times RB'}{RB + RB'} = \frac{160 \text{ k}\Omega \times 10 \text{ k}\Omega}{160 \text{ k}\Omega + 10 \text{ k}\Omega} = 9.4 \text{ k}\Omega$$

$$r_{i1} = \frac{R_{ib} \times RP}{R_{ib} + RP} = \frac{4.5 \text{ k}\Omega \times 9.4 \text{ k}\Omega}{4.5 \text{ k}\Omega + 9.4 \text{ k}\Omega} = 3 \text{ k}\Omega$$

Thus the source is looking into a load that has a total resistance of about 3 kΩ. The second stage is treated similarly. $I_E \cong 1.65$ mA and $r_e \cong 15.7$ Ω.

$$R_{ib} = (\beta + 1)(r_e + RE) = 36 \times 115.7 = 4.2 \text{ k}\Omega$$

$$RP = \frac{RB \times RB'}{RB + RB'} = \frac{100 \text{ k}\Omega \times 10 \text{ k}\Omega}{100 \text{ k}\Omega + 10 \text{ k}\Omega} = 9.1 \text{ k}\Omega$$

$$r_{i2} = \frac{R_{ib} \times RP}{R_{ib} + RP} = \frac{4.2 \text{ k}\Omega \times 9.1 \text{ k}\Omega}{4.2 \text{ k}\Omega + 9.1 \text{ k}\Omega} = 2.9 \text{ k}\Omega$$

With the input resistance of each transistor determined, we can now begin the analysis of the large-signal response, the voltage gain, etc.

We shall first determine the voltage gain of stage 1. It is very important to realize that the total load resistance for the signal is the shunt resistance of RL *and* the r_i of the following stage. That is, the total ac load for the input transistor is the 8.2-kΩ load resistor in parallel with the r_i of the second stage. That this is true is evident from the fact that signal current in the collector circuit must flow in both paths. We use the symbol r_l to designate the ac load, and this allows us to differentiate from the dc load RL.

$$r_{l1} = \frac{RL_1 \times r_{i2}}{RL_1 + r_{i2}} = \frac{8.2 \text{ k}\Omega \times 2.9 \text{ k}\Omega}{8.2 \text{ k}\Omega + 2.9 \text{ k}\Omega} = 2.14 \text{ k}\Omega$$

Now that the effective ac load has been determined, we can find the voltage gain of the input circuit.

$$A_{v1} = \frac{r_{l1}}{r_e + RE_{\text{unbypassed}}} = \frac{2.14 \text{ k}\Omega}{126} \cong 17$$

The voltage gain for the second stage is determined in like manner. (Since we do not know the ultimate load on the second stage, we must use the 10-kΩ potentiometer as the load in parallel with RL_2.)

$$r_{l2} = \frac{RL \times 10 \text{ k}\Omega}{RL + 10 \text{ k}\Omega} = \frac{6.8 \text{ k}\Omega \times 10 \text{ k}\Omega}{6.8 \text{ k}\Omega + 10 \text{ k}\Omega} = 4 \text{ k}\Omega$$

$$A_{v2} = \frac{r_{l2}}{r_e + RE_{\text{unbypassed}}} = \frac{4 \text{ k}\Omega}{115.7} = 34.6$$

The overall voltage gain (with the output unloaded) is the product of the two individual stage gains.

$$A_v = A_{v1} \times A_{v2} = 17 \times 34.6 \cong 588$$

We can now find the large-signal response of the circuit without using the characteristic curves. Keeping in mind that we are only talking about the signal conditions, the effective load in the collector circuit of the input transistor is r_{l1}. Now, the quiescent current through the 1-kΩ emitter resistor is about 1 mA; so the capacitor CE_1 is charged to about 1 V. Because of the signal conditions, this must be considered as a dc source in the circuit.

First, we must determine the maximum current that can flow in the collector-emitter circuit under signal conditions. Mentally replacing RL (8.2 kΩ) with r_l (2.14 kΩ), the circuit can be simplified as shown in Fig. 12-7b. Note that the two emitter resistors are transposed in their relative positions. This is done to simplify the description, and does not affect the calculation of the maximum current. The 2.14-kΩ resistance and the 100-Ω resistance will have 19 V across them if the transistor is driven rapidly just to the edge of saturation. This is the condition where maximum current will flow.

$$-I_{C(max)} = \frac{-19}{2.14 \text{ k}\Omega + 0.1 \text{ k}\Omega} = \frac{-19}{2.24 \text{ k}\Omega} = -8.48 \text{ mA} \cong -8.5 \text{ mA}$$

The maximum drop across the load, then, is r_l multiplied by I_{max}.

$$E_{rl(max)} = I_{max} \times r_l = 8.5 \text{ mA} \times 2.24 \text{ k}\Omega \cong 19 \text{ V}$$

Thus the collector voltage $-V_C$, referred to ground, can swing from -20 V if the transistor is in cutoff and to -1 V if the transistor is driven to saturation. This is, of course, a 19-V excursion.

Stage 2 can be analyzed in the same manner. Replacing RL with r_{l2}, as was done above, will allow us to determine the large-signal response. The emitter current, we recall, is -1.65 mA; so the drop across the 1-kΩ emitter resistor is 1.65 V. This leaves $-20 + 1.65 = -18.35$ V across the other resistors if the transistor is saturated.

$-I_{C(max)}$, then, is equal to the voltage across the resistors, divided by the value of the resistors.

$$-I_{C(max)} = \frac{-18.35}{4.1 \text{ k}\Omega} = -4.5 \text{ mA}$$

Again, the maximum drop across the load, r_{l2}, is $I_{C(max)}$ multiplied by r_{l2}.

$$E_{RL(max)} = |I_{C(max)}| \times r_{l2} = 4.5 \text{ mA} \times 4 \text{ k}\Omega = 17.9 \text{ V}$$

The collector voltage V_C, referred to ground, can swing from -20 to -2.1 V. This is a 17.9-V excursion.

Finally, the frequency response can be approximated for this circuit. The published value of the alpha cutoff frequency is in the vicinity of 2 MHz.

$$f_\beta = \frac{f_\alpha}{\beta} = \frac{2 \text{ MHz}}{35} = 57 \text{ kHz}$$

and

$$f_2 = f_\beta + \frac{f_\alpha}{A_v} = 57 \text{ kHz} + \frac{2 \text{ MHz}}{17} \cong 175 \text{ kHz}$$

The amplifier is therefore capable of amplifying frequencies somewhat less than 175 kHz.

The low-frequency response of each circuit can be determined individually. C_{c1} helps determine the low-frequency limit of the input circuit.

$$f_1 = \frac{1}{2\pi X_C C} \qquad (\text{where } X_C = r_i)$$

$$f_{1(1)} = \frac{1}{6.28 \times 3 \text{ k}\Omega \times 5 \text{ }\mu\text{F}} = 10.6 \cong 11 \text{ Hz}$$

The emitter-bypass capacitor CE_1 also affects the input low-frequency response.

$$f_{1(2)} = \frac{1}{2\pi X_C C} = \frac{1}{6.28 \times 1 \text{ k}\Omega \times 200 \text{ }\mu\text{F}} = 0.796 \cong 1 \text{ Hz}$$

The best low-frequency response of the input stage is the larger of these two figures, and is therefore about 11 Hz.

The low-frequency response of the second stage is determined the same way.

$$f_{1(3)} = \frac{1}{2\pi X_C C} = \frac{1}{6.28 \times 2.9 \text{ k}\Omega \times 10\mu\text{F}} = 5.49 \cong 5.5 \text{ Hz}$$

Since CE_2 and its associated resistor are the same values as in the input circuit, its limit $f_{1(4)}$ is the same, or about 1 Hz.

Lastly, the output coupling capacitor and the 10-kΩ resistor will affect the low-frequency response. We find that the limit $f_{1(5)}$ set by these components is on the order of 2 Hz, and this is found in the same way as in the last several examples.

Now, the overall response of the amplifier is slightly higher than the largest number calculated above. This can be determined as follows.

$$f_1 = \sqrt{(f_{1(1)})^2 + (f_{1(2)})^2 + \cdots + (f_{1(5)})^2}$$
$$= \sqrt{121 + 1 + 1 + 30.25 + 4} = \sqrt{156.25}$$
$$\cong 12.5 \text{ Hz}$$

If another transistor were used to amplify the signal further, this would load the output coupling capacitor, and the response might reasonably be expected to be 20 to 30 Hz.

QUESTIONS AND PROBLEMS

12-1 In the circuit shown, if $\beta = 100$, determine
(a) the voltage gain
(b) the total ac input resistance
(c) gamma

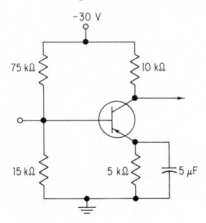

12-2 In the circuit shown, what is the approximate low-frequency limit?

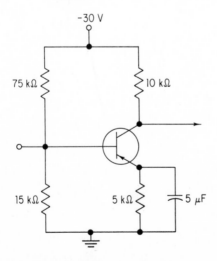

12-3 In the circuit shown, determine
 (a) the large-signal voltage limits of V_C
 (b) the required input voltage to just cause maximum collector excursion

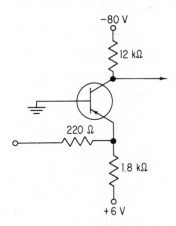

12-4 In the circuit shown, what are the input resistance r_i, the voltage gain, and the dc emitter current?

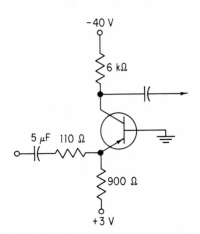

12-5 Using the circuit shown, determine the total signal input resistance. Assume midband conditions. Transistor parameters are given:

$$V_{CE(\text{max})} = 30 \text{ V}; \ I_{C(\text{max})} = 300 \text{ mA}; \ P_{C(\text{max})} = 250 \text{ mW}; \ \alpha = 0.99.$$

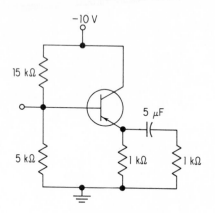

12-6 Refer to Fig. 12-1. Determine the dc conditions for the circuit if RB is changed in value to 500 kΩ.

12-7 Refer to Fig. 12-3. Change R_c to 500 Ω. What is the new voltage gain?

12-8 Refer to Fig. 12-2. Change RB to 22 kΩ. Determine the dc conditions of this circuit. Is A_v changed significantly?

12-9 Refer to Fig. 12-7. Change the load resistor of the output transistor from 6.8 to 4.7 kΩ. What is the voltage gain for the output stage?

12-10 Refer to Fig. 12-7. Discuss the effect on the circuit response of changing C_{C1} to 0.1 μF.

12-11 Refer to Fig. 12-1. Change the load resistor RL from 5000 to 2500 Ω. Determine the value of R_{CB}.

12-12 Refer to Question 12-11. Find the value of I_C.

12-13 Refer to Question 12-11. Find the value of V_C.

12-14 Refer to Question 12-11. Find the value of I_B.

12-15 Refer to Question 12-11. Find the value of gamma.

12-16 Refer to Question 12-11. Find the value of r_e.

12-17   Refer to Question 12-11. Find the value of A_v.

12-18   Refer to Question 12-11. Find the value of R_{BE}.

12-19   Refer to Fig. 12-2. Change RB to 18,000 Ω and RB' to 6000 Ω. Determine the value of R_{CB}.

12-20   Refer to Question 12-19. Find the value of I_C.

12-21   Refer to Question 12-19. Find the value of V_C.

12-22   Refer to Question 12-19. Find the value of I_B.

12-23   Refer to Question 12-19. Find the value of gamma.

12-24   Refer to Question 12-19. Find the value of R_{ib}.

12-25   Refer to Question 12-19. Find the value of A_v.

12-26   Refer to Question 12-19. Find the value of r_i.

12-27   Refer to Fig. 12-2. Change the value of RB to 27,000 Ω and RB' to 9000 Ω. Find the value of R_{CB}.

12-28   Refer to Question 12-27. Determine the value of I_C.

12-29   Refer to Question 12-27. Find the value of V_C.

12-30   Refer to Question 12-27. Find the value of I_B.

12-31   Refer to Question 12-27. Find the value of gamma.

12-32   Refer to Question 12-27. Find the value of A_v.

12-33   Refer to Question 12-27. Find the value of r_i.

CHAPTER 13 TRANSISTOR OSCILLATORS

An oscillator is a circuit that produces an ac output by virtue of providing its own input. There are many different kinds of oscillator circuits, and each is dependent upon the requirement that there be amplification somewhere in the circuit. Hence most oscillators are simply amplifiers, with a portion of the output applied to the input as feedback. A block diagram is shown in Fig. 13-1 that is helpful in understanding these interesting circuits.

The output of the amplifier is seen to be presented back to its input through the feedback circuit. The main requirement here is to ensure that the feedback signal is applied in phase with any signal already at the input. Thus the feedback circuit must provide some phase shift, usually. If the amplifier circuit is a common-emitter circuit, which provides a 180° phase shift, the feedback circuit must provide an additional 180° shift. On the other hand, if the amplifier is a common-base circuit, the feedback circuit does not need to shift the phase at all, since the output (at the collector) and input (at the emitter) are in phase already.

Another important criterion is the requirement that the total loop gain, $A \times B$, must be greater than 1. The gain of the feedback circuit is usually less than 1; so the gain of the amplifier must be appreciable to overcome this loss. This is usually written as gain, $A \times$

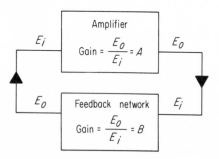

FIG. 13-1. Block diagram of an oscillator.

$B > 1$. If power gain is being considered, this relation is called the *Barkhausen criterion*. Usually, however, it is enough simply to say that the requirement is met if the voltage gain is greater than 1. In a practical circuit, an active device, such as a transistor, provides the amplification, while passive components, such as resistors, capacitors, and inductors, provide the feedback network. As mentioned, the feedback network may or may not provide some degree of phase shift, but almost always provides some loss.

13-1 *RC*-COUPLED PHASE-SHIFT OSCILLATOR

A simple oscillator is shown in Fig. 13-2. The amplifier is a simple self-bias circuit, and the voltage between base and collector is 180° out of phase. The purpose of the high-pass network is to provide an additional 180° of phase shift, while at the same time not attenuating the output of the amplifier by too great a factor. Any signal appearing at the base, due perhaps to noise or any irregularity, is amplified by the transistor and presented to the phase-shifting network. At some frequency the network will provide a full 180° shift, and at this frequency the output, fed back to the base, will have been shifted by a full 360°. Hence this new signal appearing at the base is in phase with, and will reinforce, the original signal.

This reinforced signal will again be reamplified and reapplied to the base; the circuit is said to be regenerating, and the transistor is alternately turned all the way on and all the way off. The output of the oscillator, taken between the collector and ground, will be a sinusoid. If the gain is somewhat greater than 1, the amplitude of the generated signal will approximate the collector supply voltage. If the gain is only slightly greater than 1, the output may be somewhat smaller in amplitude.

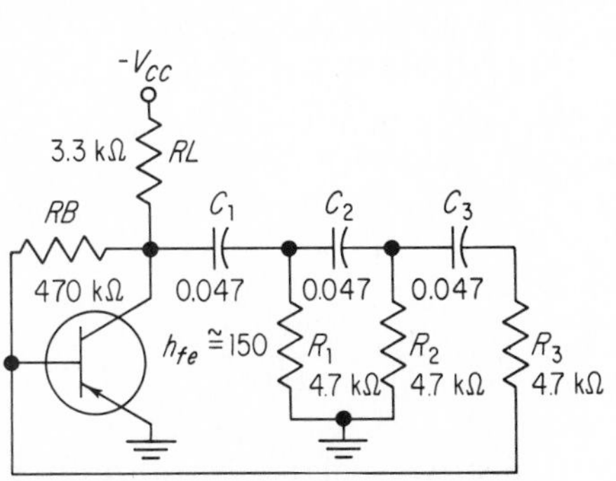

FIG. 13-2. Phase-shift oscillator.

The frequency at which the individual RC circuits shift the phase by 60° will be the frequency of oscillation. An approximate relation that allows this frequency to be determined is shown as follows:

$$f_{osc} = \frac{1}{2\pi \sqrt{Y + Z}}$$

where $Y = 6(R^2C^2)$
$Z = 4(RR_LC^2)$
$R = 4.7 \text{ k}\Omega$
$C = 0.047 \text{ } \mu\text{F}$
$R_L = 3.3 \text{ k}\Omega$

To calculate the frequency of oscillation for the circuit shown in Fig. 13-2 the proper values are inserted in the formula and the expression then evaluated.

$Y = 6(R^2C^2) = 6(4.7 \text{ k}\Omega)^2(0.047 \text{ } \mu\text{F})^2$
$= 6(22.1 \times 10^6)(0.00221 \times 10^{-12})$
$= 0.293 \times 10^{-6}$

$Z = 4(R \times R_L \times C^2) = 4(4.7 \text{ k}\Omega \times 3.3 \text{ k}\Omega \times 0.00221 \times 10^{-12})$
$= 0.1371 \times 10^{-6}$

$$f_{osc} = \frac{1}{2\pi \sqrt{Y + Z}} = \frac{1}{2\pi \sqrt{(0.293 \times 10^{-6}) + (0.1371 \times 10^{-6})}}$$

$$\cong \frac{1}{6.28 \times \sqrt{0.4301 \times 10^{-6}}} = \frac{1}{6.28 \times 6.56 \times 10^{-4}}$$

$$= \frac{1}{4.118 \times 10^{-3}} = 243 \text{ Hz}$$

This oscillator, then, would operate at a frequency of about 243 Hz.

The biasing of this circuit is very critical, for to get the maximum peak-to-peak output with a minimum of distortion, the transistor is biased near the center of the ac load line. This particular circuit is biased near the center of the dc load line, and in most similar circuits this is sufficient.

$$R_{CB} = \frac{RB}{h_{FE}} = \frac{470 \text{ k}\Omega}{150} = 3.1 \text{ k}\Omega$$

Note that this is very nearly equal to the load resistor, which indicates the center of the load line. If the transistor were biased nearer to either extreme of the load line, the danger exists that the output wave-form, taken from the collector, would be severely clipped. In many applications this would be most undesirable.

This circuit is primarily useful as a low-frequency oscillator, and is often used to generate audio frequencies. To allow the oscillator to be made variable, a potentiometer can be added in place of, for instance, R_2, and this will change the phase shift by some amount, and therefore the frequency of oscillation. To prevent the oscillator from stopping, a 1-kΩ resistor should be placed in series with the potentiometer. This will prevent the resistance in the high-pass network from being shorted to ground.

13-2 THE ARMSTRONG OSCILLATOR

Many oscillator circuits use various combinations of inductors and capacitors to help produce oscillations. Probably the simplest of these circuits is the Armstrong, or tickler, oscillator. Such a circuit is shown in Fig. 13-3. When a tank circuit, as shown in the figure, is pulsed with a voltage, it will ring, or oscillate, for a brief period of time. This effect produces a damped waveform, as shown in Fig. 13-4. The loss of electrical energy as indicated is due to the resistance encountered by the current. The purpose of the oscillator circuit is to replenish the energy periodically so as to encourage continuous oscillation. In the circuit shown, the transistor conducts only for a small part of each cycle, and collector current provides for the additional energy supplied once each cycle.

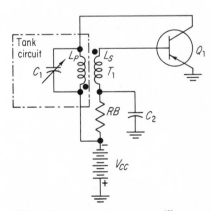

FIG. 13-3. Armstrong oscillator.

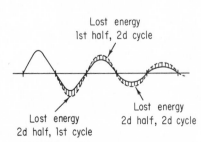

FIG. 13-4. Waveform of damped oscillations.

In this circuit configuration the transistor itself provides 180° of phase shift, and the transformer yields the remainder, thus effecting complete 360° phase shift. If distortion is not too great a factor, this type of circuit is often operated class C, and the transistor then acts simply as a switch, turning on briefly once each cycle to replenish the tank circuit. The collector-current waveform, then, is a large pulse that is zero most of the time. However, the voltage across the tank is, of course, a very good sine wave, since the reactive components supply energy during the complete cycle. By adding a third winding to the transformer a means is provided to extract a sine wave from the circuit. This, however, would load the tank circuit and cause the frequency to be changed.

The frequency of oscillation for this circuit (unloaded) is determined approximately by the constants of the tank circuit.

$$f_{osc} = \frac{1}{2\pi\sqrt{LC}}$$

This circuit configuration is not particularly stable as far as frequency is concerned; hence it is seldom used in practical cases. A closely related circuit that has better frequency stability is shown in the following section.

13-3 THE HARTLEY OSCILLATOR

The Hartley circuit is very similar to the Armstrong oscillator, the primary difference being in the coil arrangement. Instead of separate

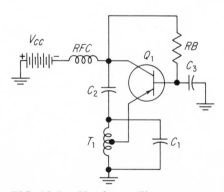

FIG. 13-5.   Hartley oscillator.

primary and secondary windings, the Hartley circuit uses a tapped coil, or autotransformer. This still allows regeneration, but somewhat simplifies the circuit arrangement. A typical circuit is shown in Fig. 13-5, where the tapped coil is indicated. T_1 and C_1 form the resonant tank circuit, with emitter current providing the feedback.

This particular circuit happens to be a common-base type, as evidenced by the capacitor C_3, holding the base at signal ground. Because the collector voltage and emitter voltage are in phase, there is no need to provide additional phase shift. The amplified collector voltage is applied to the emitter through C_2, and this signal turns the transistor on at the proper moment. Emitter current flows through the lower part of T_1 and replenishes the tank current. The inductance of T_1 and the capacitance of C_1 determine the resonant frequency of the oscillator.

13-4   THE COLPITTS OSCILLATOR

When the capacitor leg of the tank circuit is tapped, rather than the inductor leg, the circuit is called the *Colpitts oscillator*, a typical example of which is shown in Fig. 13-6. This circuit is operating in the common-base configuration, and can be seen to be biased in the familiar universal circuit. C_1 and C_2 form a capacitive voltage divider, feeding back some part of the total tank voltage to the emitter. The ratio of C_1 to C_2 determines the amount of feedback presented to the emitter.

The bias point for this circuit is rather easily determined. The

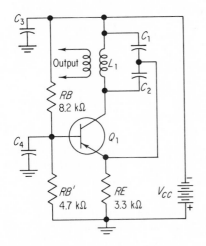

FIG.   13-6.   Colpitts   oscillator.

ERCA equation for a circuit such as this is shown. With the values inserted, the circuit can be analyzed for its quiescent operating point.

$$R_{CB} \text{ is to } RB \text{ as } RE \text{ is to } RB'$$

$$R_{CB} \times RB' = RB \times RE$$

$$R_{CB} = \frac{RB \times RE}{RB'} = \frac{8.2 \text{ k}\Omega \times 3.3 \text{ k}\Omega}{4.7 \text{ k}\Omega} = 5.76 \text{ k}\Omega$$

$$\gamma = \frac{R_{CB}}{R_{CB} + RE} = \frac{5.76 \text{ k}\Omega}{5.76 \text{ k}\Omega + 3.3 \text{ k}\Omega} = 0.64$$

The transistor is biased slightly below the midpoint of the load line. If R_{CB} were made equal to RE, it would be biased at the very center of the load line. This is true because there is no significant dc resistance in the collector lead.

The frequency of oscillation depends upon the values of L_1, C_1, and C_2. For instance, if L_1 is 200 μH, C_1 is 1500 $\mu\mu$F, and C_2 is 800 $\mu\mu$F, what is the resonant frequency, f_{osc}?

$$L = 0.2 \text{ mH} = 200 \ \mu\text{H}$$

$$C_{total} = \frac{C_1 C_2}{C_1 + C_2} = \frac{1.2 \times 10^{-18}}{2.3 \times 10^{-9}} = 522 \ \mu\mu\text{F}$$

$$f_{osc} = \frac{1}{2\pi \sqrt{LC}} = \frac{1}{6.28 \sqrt{(2.0 \times 10^{-4})(5.22 \times 10^{-10})}}$$

$$= \frac{1}{6.28 \sqrt{10.44 \times 10^{-14}}} = \frac{1}{6.28 \times 3.23 \times 10^{-7}}$$

$$= \frac{1}{20.3 \times 10^{-7}} \cong 0.0493 \times 10^7 = 493 \text{ kHz}$$

13-5 PUSH-PULL COLPITTS OSCILLATOR

The circuit shown in Fig. 13-7 is a balanced, or push-pull, oscillator of the Colpitts type. It is most useful to provide a sinusoidal output waveform that contains fewer harmonic distortion components than is obtained from a single-ended circuit. The capacitive voltage divider consisting of C_1, C_2, and C_3 is symmetrically arranged in the collector circuit. Along with the inductance of T_1, the capacitors form a resonant circuit that determines the frequency of operation for the

oscillator. Further, they tap a portion of the tank energy and deliver
it to each transistor base so as to effect regenerative feedback. Hence,
as long as Q_1 and Q_2 are biased in the active region and the overall
circuit gain is high, the circuit will function as an oscillator, producing
a sine wave at a frequency determined by the total values of L and C.

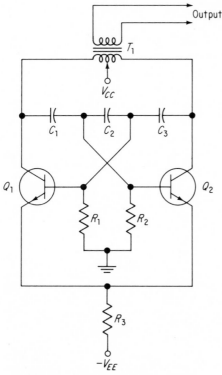

FIG.   13-7.   Push-pull   Colpitts
oscillator.

To describe the circuit action in general terms, assume that the
collector of Q_1 at one point in time is negative-going. Q_1 is therefore
in the process of turning on harder. This negative-going voltage at
the collector of Q_1 is also seen at the base of Q_2 through C_1. A more
negative voltage at the base of Q_2 turns it off further; hence its collector voltage rises in the positive direction. This positive-going voltage is, through C_3, seen at the base of Q_1, turning it further on. This
regenerative action continues until the loop gain is reduced to a value
too low to sustain further action in this direction. Now, the collector
of Q_1 begins to rise in the positive direction since there is no more

energy, via C_3, applied to its base to sustain a fully-on condition. Q_1 begins to turn off, and its collector starts to rise in the positive direction, turning Q_2 on somewhat. This action continues until the opposite extreme is reached and the loop gain is again reduced to perhaps unity. As long as power is applied to the circuit, it will continue to oscillate, providing the transistors are correctly biased and the loop gain is sufficiently high. The output waveform is determined by the resonant tank circuit in the collector circuits. This will, of course, be a sinusoidal waveshape if the transistors are not driven so hard as to produce significant distortion.

The circuit is biased rather simply. R_1 and R_2 are chosen to be low enough in value to ensure that the bases of the transistors will be essentially at ground potential, but high enough to allow the signal voltage to be developed across them. A typical value for this might be 1 kΩ. R_3 is then given a value to provide each transistor with the correct emitter current. Assume that each transistor is to have an

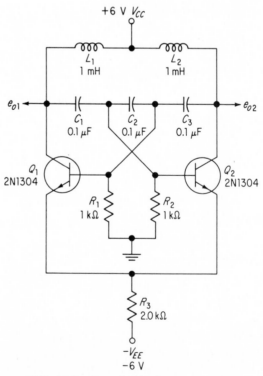

FIG. 13-8.   Practical Colpitts oscillator.

emitter current of 3 mA. The total quiescent emitter current is therefore 6 mA. Since each base is essentially at ground potential, so also
are the emitters, if we ignore V_{BE}. The value of R_3 is then found as
follows:

$$R_3 = \frac{E_{RE}}{I_{total}} = \frac{6}{0.006} = 1000 \ \Omega$$

Figure 13-8 illustrates a similar circuit that can readily be bread-
boarded in the shop from available components. Such circuit characteristics as quiescent voltages and currents, frequency of operation,
and waveshape can be determined in the classroom and later verified
on the bench.

13-6 RELAXATION OSCILLATORS

A relaxation oscillator uses one or more capacitors to cause oscillation
and, excepting the phase-shift oscillator previously mentioned, usually produces a square-wave output. This interesting circuit is one of
a class of *RC* oscillators called *astable multivibrators.* The name is derived from the fact that there is no stable state, and the circuit oscil-
lates because it is continually hunting for a stable state. As in any
conventional oscillator, there is regeneration. In this case the phase
shift necessary to produce regeneration is provided by an additional
transistor. Hence the circuit generally encountered uses two transistors, one to give 180° of phase shift, and the other to yield the remaining 180°. As usually designed, both transistors provide gain,
and thus the loop gain is very high.

Astable Multivibrator

A relaxation oscillator may be defined as one in which the waveform
changes abruptly and periodically to produce a nonsinusoidal waveform, usually a square wave. The circuit operates because of the time
delay of capacitive charge or discharge through a resistor.

A typical astable multivibrator circuit is shown in Fig. 13-9. Note
the closed loop; it is this that provides the regeneration and gain
necessary to cause oscillation. This feedback loop is emphasized in
Fig. 13-10, where the same circuit is redrawn somewhat. The output
of Q_1 is fed to the input of Q_2, while the output of Q_2 is returned to
the input of Q_1. This forms the closed loop, and the 360° phase shift
occurs because each transistor provides 180°.

Note, also, that both transistors are operated in the common-emitter configuration; hence the loop voltage gain can be made very high. The smallest change at any point in the circuit is amplified, inverted, reamplified, and reinverted. It thus appears at the starting point greatly amplified, as suggested by the polarity signs in the figure.

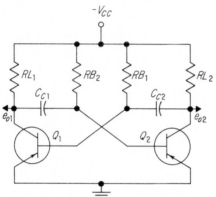

FIG. 13-9. Astable (free-running) multivibrator.

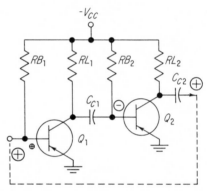

FIG. 13-10. Regenerative closed loop.

To appreciate how the circuit works, a simplified drawing is given in Fig. 13-11, where the transistors are replaced by simple switches. One stipulation must be made to cause the simplified circuit to perform like the actual circuit. If switch 1 is open, switch 2 *must* be closed. If switch 1 is closed, switch 2 *must* be open.

As drawn, switch 1 is closed, and switch 2 is open. The waveforms shown indicate the conditions at A and B as the switches are alternately transferred. Initially, with switch 1 closed, a large current flows from $-V_{CC}$ through the load resistor and out to ground. The full $-V_{CC}$ is dropped across the resistor, and point B is therefore at ground, or 0 V. At the same time, point A is at $-V_{CC}$ since, with switch 2 open, there is no current flow through the resistor, and therefore no voltage drop across it.

When the switches transfer, point B will rise to $-V_{CC}$, while point A will fall to ground, or 0 V. As the switches are continually transferred, the output from either wire, A or B, is a square wave, each being 180° out of phase with the other.

The multivibrator circuit shown in Fig. 13-9 operates in the same manner. One transistor conducts, while the other is cut off. Then the situation reverses, automatically, at a frequency determined by the time constants in the circuit. When power is applied to this circuit, the transistors alternately conduct and cut off, and will produce a train of square waves for as long as power is applied.

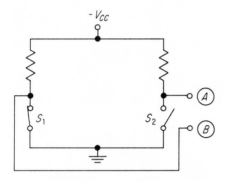

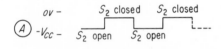

FIG. 13-11. Simplified equivalent circuit of multivibrator with waveforms.

13-7 THE VOLTAGE-CONTROLLED OSCILLATOR

A voltage-controlled oscillator (VCO) is one in which the frequency of operation is caused to vary as a function of a varying dc control voltage. Such a circuit is useful in a number of applications, among which are servofeedback circuits, automatic frequency control, automatic phase control, and pulse-interval modulators.

One such circuit, shown in Fig. 13-12, is essentially an *RC*-coupled multivibrator that is modified to accept a dc control voltage. The control voltage is introduced in a manner such as to influence the charge and/or discharge of the timing capacitors. This, of course, will change the frequency of oscillation.

In the circuit shown, the oscillator is designed to operate at a frequency of 444 Hz with the dc control voltage open-circuited. With a small dc voltage supplied at the control input, the frequency changes. A more positive input results in a higher frequency, while a change in the negative direction yields a lower frequency.

The network shown associated with R_6, R_7, and R_8 is useful to demonstrate the frequency variation with a change in control voltage.

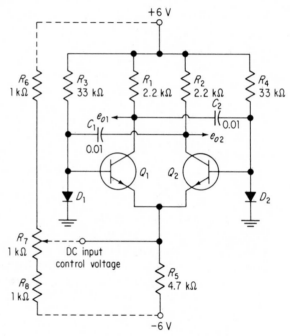

FIG. 13-12.   Voltage-controlled oscillator.

Simply observing the waveform at either collector while varying the potentiometer R_7 will graphically demonstrate the main feature of this circuit. The lowest frequency at which it will function is approximately 225 Hz, while the highest is 28 kHz, a better than 100-to-1 range.

A basic circuit evaluation reveals that Q_1 and Q_2 form a fundamental astable multivibrator. With no dc control voltage applied, the natural frequency of oscillation is determined by R_3, R_4 and C_1, C_2. The conventional feedback loop is easily recognized: the output of Q_1 is fed to the input of Q_2, which is amplified and inverted and is in turn fed to the input of Q_1. Q_1 amplifies and inverts the signal, and its output is directed to the input of Q_2, etc.

One feature of this circuit is somewhat different from a conventional multivibrator. Note D_1 and D_2; these are essential to the proper operation of the circuit. During each respective half-cycle, it is necessary that C_1 and C_2 recharge to the same value of voltage regardless of the value of control voltage, and the two diodes provide this action. If they are removed from the circuit, the oscillator will function, but the dc control voltage will have little or no effect on the frequency.

13-8 CRYSTAL-CONTROLLED OSCILLATOR

Crystal-controlled oscillators have many applications. Wherever the need exists to produce a waveform that is much more stable than a simple LC combination will produce, the crystal oscillator is used. Figure 13-13 illustrates such a circuit. In this instance, the crystal used is a 27-MHz (citizens' band) crystal with ±0.005 percent accuracy. One reason for using such a crystal in this example is that it is relatively simple to obtain and at a minimum cost. Furthermore, any low-cost CB receiver can be used to monitor the output of the oscillator to verify that it is operating properly.

The oscillator is biased in the familiar universal circuit with the

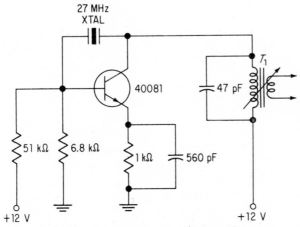

FIG. 13-13. Crystal-controlled oscillator.

voltage divider consisting of the 51-kΩ and the 6.8-kΩ resistors providing 1.3 V at the base. Emitter current is limited by the 1-kΩ resistor. To maintain large voltage gain, this resistor is bypassed with a 560-pF capacitor. The resonant tank circuit in the collector lead is tuned for resonance at 27 MHz. The transistor used is an RCA 40081, a transistor especially designed for such operation. It is capable of producing about 1 or 1.5 W of RF into a suitable load.

This particular circuit makes an excellent laboratory project. The output is sufficient to cause RF quieting on a nearby receiver, but not sufficient to cause illegal transmission. A short length of wire attached to the collector will aid in receiving the output. The RF output can be modulated very simply by connecting two wires, originating across the volume control of a *battery-operated* AM receiver, and connected across the 1-kΩ emitter resistor.

QUESTIONS AND PROBLEMS

13-1 Select the correct answer. The oscillator circuit that uses a tapped inductance in the tuned circuit is
(*a*) the Colpitts   (*b*) the multivibrator
(*c*) the Hartley   (*d*) the Armstrong

13-2 Select the correct answer. The oscillator circuit that uses "tapped" capacitors in the tuned circuit is
(*a*) the Colpitts   (*b*) the multivibrator
(*c*) the Hartley   (*d*) the Armstrong

13-3 What is the resonant frequency in the circuit shown?

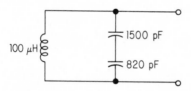

13-4 What is the illustrated waveform called?

13-5 Refer to Fig. 13-8. Determine the operating frequency if L_1 and L_2 are each 5 mH and C_1, C_2, and C_3 are each 0.47 μF.

13-6 Refer to Fig. 13-8. Determine the operating frequency if L_1 and L_2 are 1 μH and C_1, C_2, and C_3 are each 0.001 μF.

13-7   Refer to Fig. 13-9. If Q_1 is on, what is Q_2?

13-8   Refer to Fig. 13-9. If Q_2 is on, what is Q_1?

13-9   Refer to Fig. 13-13. If the crystal is cut to exactly 27 MHz, what must be the inductance of T_1 to resonate at 27 MHz?

13-10 Refer to Fig. 13-13. If L_1 is equal to 50 μH and the shunt capacitor is 47 pF, what must be the crystal frequency?

CHAPTER 14 LINEAR-CIRCUIT EXAMPLES

14-1 CLASSES OF LINEAR AMPLIFIERS

Linear amplifiers are classed according to their mode of operation. There are many ways of doing this, some of which are shown in the following list. When describing an amplifier, we might use any of these terms.

1. *Power amplifier*   Output classification
   Voltage amplifier   Output classification
2. *Small-signal amplifier*   Input classification
   Large-signal amplifier   Input classification
3. *RF amplifier*   Frequency classification
   IF amplifier   Frequency classification
   AF amplifier   Frequency classification
4. *Class A amplifier*   Bias classification
   Class B amplifier   Bias classification
   Class C amplifier   Bias classification
   Class AB amplifier   Bias classification
5. *Single-ended*   Circuit configuration
   Push-pull   Circuit configuration
   Parallel   Circuit configuration

Each of these terms helps to categorize a particular amplifier as to its "mode" of operation, which simply means the way it operates, according to a predetermined set of values. Each group has been given a name that describes the type of classification. For instance, we might describe a circuit as a small-signal class-A voltage amplifier operating in the audio range. From this description, one could very nearly draw the circuit in question, except for component values.

Whether a circuit is class A, class B, or another class is determined by the limits of the circuit. In this case, the limits of a circuit are determined by the bias arrangement afforded. In other words, the bias of the circuit sets the limits, and so also sets the class of operation.

Class A Amplifier

The class A amplifier is a circuit that is intended for use as a linear amplifier. That is, the circuit must produce so little distortion that the output is, so far as possible, an exact, although amplified, replica of the input. The signal must never drive the transistor to either circuit limit (cutoff or saturation), for if this occurs, there will be severe distortion. Thus the transistor must be biased somewhere in the center of its characteristics, so that it can be allowed to swing either side of the quiescent operation point. This is shown graphically in Fig. 14-1.

In Fig. 14-1 are shown the characteristic curves of a transistor with a dc load line drawn on them. The two ends of the load line, at

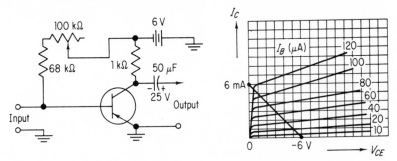

FIG. 14-1.   Simple class A audio amplifier.

−6 V and 6 mA, are the limits of this particular circuit. If it is to be operated as a class A amplifier, the transistor must never reach either limit. In other words, it must always remain in the "on" condition. Thus the transistor is usually biased toward the center of the load line for class A operation.

One other factor is of importance concerning the class A amplifier. The amplitude of the input signal must be small enough to ensure that neither limit is ever reached. The maximum input signal, peak to peak, is usually specified for a given amplifier, so that this may be verified. If the input signal is a very small one, it does not make any difference where on the load line the transistor is biased as far as the limits are concerned.

Typical waveforms for this circuit are illustrated in Fig. 14-2. They show that collector current flows all the time, and that at no time does it go to either cutoff or to saturation.

The circuit in Fig. 14-1 is a typical example of a simple class A voltage amplifier. The biasing resistors, R_1 and R_2, are set to turn the transistor partly on, so that the signal at the input can add to, or subtract from, the normal quiescent current. Since the transistor's beta

determines the amount of collector current that will flow, and since
no two transistors have exactly the same beta, the circuit must be read-
justed for each transistor placed in the circuit. This is the purpose
of the variable resistor in the base lead. To ensure that the transistor
is set to the approximate center of the characteristics, the potentiom-

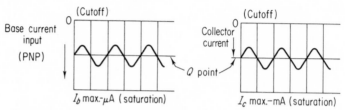

FIG. 14-2.   Collector and base currents, class A amplifier.

eter is adjusted until the collector voltage is about one-half V_{CC}, or,
in this example, to about -3 V.

Class B Amplifier

The class B amplifier is biased differently from the class A type. The
transistor used in a class B circuit is biased so that it will *not* produce
an exact replica of the input signal. A single transistor used this way
badly distorts the input signal, as can be seen in Fig. 14-3. Note that
output current flows only during one-half of the input cycle. The
transistor is biased exactly at cutoff, and during one-half of the input
cycle the transistor is turned further off. However, during the other
half-cycle the transistor is turned on *by the signal,* and collector cur-

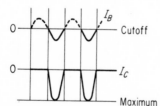

FIG. 14-3.   Class B amplifier waveforms.

rent flows. It may or may not increase to the maximum value as shown
in the figure.

   In the PNP case illustrated, when the base is caused to go more
positive, the transistor is turned off. Since it was already turned off,
no change occurs at the collector. When the signal causes the base to

go more negative, the transistor is turned on and collector current begins to flow. This half of the input cycle is amplified linearly and is a fairly good replica of the corresponding half of the input cycle.

If in special cases the class B amplifier must yield symmetrical outputs, it is possible to use two transistors. Now each provides one-

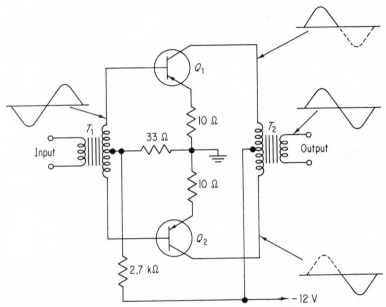

FIG. 14-4.   Class B push-pull power amplifier.

half of the total signal. By combining the two halves, the full undistorted output can be obtained. When used this way, the circuit is called a *push-pull* circuit, for as one transistor is "pushing" (turning on), the other is "pulling" (turning off).

A typical circuit is shown in Fig. 14-4. Each transistor is quiescently biased close to cutoff. If the signal is such as to cause the base of Q_1 to go more negative, more current flows in its collector circuit and the first half cycle appears at the output, which is the secondary of the T_2 transformer. As the input swings the other side of the base line, Q_2 now comes on and provides the other half-cycle. Thus the complete cycle is evident at the secondary of the output transformer. This kind of circuit is commonly used for large power output, and therefore the transformers are used for proper impedance matching to ensure maximum power transfer.

Class C Amplifier

A class C amplifier is somewhat similar to a class B circuit, except that it is biased so that collector current flows for much less than one-half cycle. This is shown in Fig. 14-5, where the fact that the output current flows only for a very short time is evident. This, of course,

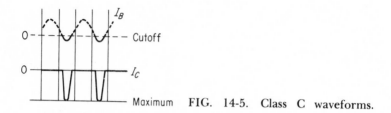

FIG. 14-5. Class C waveforms.

produces severe distortion, since the transistor is on only about one-third of the input cycle. The circuit would be used only in cases where this could be tolerated. One example is an RF amplifier, where a resonant tank circuit in the output restores the waveshape. In this instance, the transistor periodically pulses the resonant circuit, and the result is a reasonably distortion-free output.

14-2 TYPICAL AUDIO AMPLIFIERS

The circuit of Fig. 14-6 is typical of the output stage of a small transistor radio. The output transistor Q_1 is transformer-coupled to

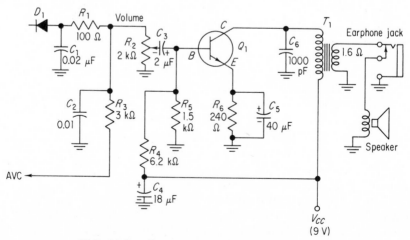

FIG. 14-6. Output stage of a transistor radio.

the speaker. This is usually the case when a single-ended output stage is used. Except for the transformer coupling, the circuit is not significantly different from the circuits we have studied thus far. The biasing network is that of the familiar universal circuit, with R_4 and R_5 forming the voltage divider that supplies the base with forward bias. The signal input arrives from D_1, which is the detector, via the volume control, R_2. The audio signal is coupled by way of C_2, the coupling capacitor, into the base of the transistor.

The dc voltages that exist at the various parts of the circuit can be determined easily by the proper use of the ERCA equations for the universal circuit. However, we shall have to take into consideration that, because of the lack of dc resistance in the collector of Q_1, the equations will have to be slightly altered. In most cases, the small transformer, T_1, will have a primary resistance of about 100 Ω; so we can consider that this is the equivalent dc resistance in the collector.

$$RB \times RE = RB' \times (RL + R_{CB})$$

$$R_{CB} = \frac{RB \times RE}{RB'} - RL = \frac{6.2 \text{ k}\Omega \times 0.24 \text{ k}\Omega}{1 \text{ k}\Omega} - 100 = 1.5 \text{ k}\Omega$$

$$I_C = \frac{V_{CC}}{R_{CB} + RE} = \frac{9}{1.5 \text{ k}\Omega + 0.24 \text{ k}\Omega} \cong 5 \text{ mA}$$

$$V_E = I_E \times RE = 0.005 \times 240 = 1.2 \text{ V}$$

$$E_{RL} = I_C \times RL = 0.005 \times 100 = 0.5 \text{ V}$$

$$V_C = 9 - 0.5 = 8.5 \text{ V}$$

The dc values of this circuit, then, are easily found.

Another example of an output stage is shown in Fig. 14-7. This is a push-pull circuit, using transformer coupling at both the input and output, similar to the circuit shown in Fig. 14-4. Transformer T_1 provides the necessary phase inversion for the bases of Q_1 and Q_2. If the base of Q_1 is driven positive, the base of Q_2 must be driven negative. This, plus the matching of impedances, is the function of T_1. T_2 must match, as closely as possible, the output impedance of the transistors to the low impedance of the speaker voice coil.

The circuit operation is quite straightforward, and the only explanation required is a word about the biasing. The two 10-Ω resistors provide current limitation to prevent excessive transistor current on large input peaks. Also, they provide a degree of temperature stabilization, as does any resistance in the emitter lead. In push-pull circuits such as this one, it is necessary to forward-bias the transistors slightly. In theory, one could allow them to remain quiescently at cutoff, since the push-pull arrangement will restore the waveform. However, transistors are notoriously nonlinear when

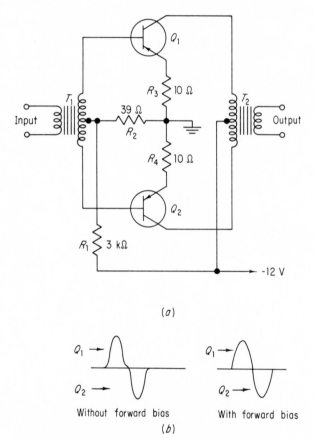

FIG. 14-7. Push-pull output stage: (*a*) without forward bias; (*b*) with forward bias.

operating near cutoff. To prevent severe crossover distortion, they are biased so that they are somewhat in the active region.

This is one purpose of R_1 and R_2. R_1 is the base resistor through which base current flows, and the small resistor, R_2, produces a voltage drop that slightly forward-biases the base-emitter junction. This works out to about 0.15 V of forward bias, and will allow an appreciable collector current to flow, preventing the crossover distortion, as suggested in Fig. 14.7*b*.

A final example of audio amplifiers is shown in Fig. 14-8. This is a complete diagram of a four-stage amplifier with push-pull output similar to the preceding example. The preamplifiers Q_1 and Q_2 are operating in the universal circuit. Each has a small unbypassed resistor in the emitter lead to make the inherent distortion less severe.

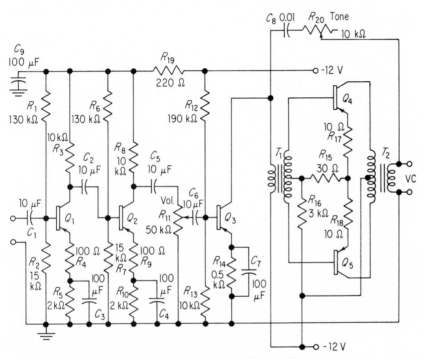

FIG. 14-8.   Complete audio amplifier.

Each of these transistors is biased slightly upward from the center of the dc load line so as to approach the center of the ac, or signal, load line. Having analyzed several very similar circuits for both the dc conditions and the signal conditions, we shall not do so at this time. It will be instructive for the reader to perform both a dc and an ac analysis of at least Q_1 and Q_2.

The overall gain of the amplifier is adjusted by the volume control R_{11}. R_{20}, along with C_8, forms a tone-control circuit that is of the high-cut type. That is, as R_{20} becomes lower in value, more of the high frequencies are fed back, out of phase, and so the gain at these higher frequencies becomes less. Thus the frequency response of the overall amplifier can be adjusted to suit the individual taste. Q_3 and T_1 provide the current drive necessary to feed the bases of Q_4 and Q_5. T_1 also acts as the phase inverter, so that the bases of the output transistors can be driven in opposite directions. Using suitable power transistors, outputs of about 1 W can be delivered to the speaker from an amplifier such as this.

An important feature of this kind of circuit is that it is usually necessary to provide "decoupling" filters, such as R_{19} and C_9, to prevent "motorboating," or low-frequency oscillations. Since all sources of power possess internal resistance, the voltage at V_{CC} will fluctuate as transistor current increases and decreases. If this is fed back to the preamplifiers *in phase*, oscillations may occur. R_{19} and C_9 act as an *RC* filter to prevent this.

14-3 GENERAL DESCRIPTION OF RF CIRCUITS

A radio-frequency (RF) amplifier differs considerably from the circuits we have discussed so far. The major difference is the ability to amplify much higher frequencies and to select certain bands of frequencies while rejecting others. Figure 14-9 illustrates in simple terms what the usual RF amplifier is expected to do. It will reject all frequencies except the band to which it is tuned.

Radio-frequency circuits can be divided into three general categories: (1) those with a fixed frequency load, (2) those with a variable frequency load, and (3) those that supply their own input (the oscillators). Most of these are represented by the RF and IF stages of a typical transistor radio. Therefore we shall use such a circuit to describe briefly these various functions.

Our main purpose at this time is to show typical uses of transistors in RF circuits, and not to explain the fundamentals of superheterodyne receivers. Many other references give excellent descrip-

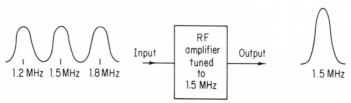

FIG. 14-9.   RF amplifier frequency rejection.

tions of typical RF circuits. However, a very brief explanation of a superheterodyne receiver follows to aid in understanding these principles. The basic circuit functions on the principle of *heterodyning*, or mixing. That is, when two signals at two separate frequencies are injected into a nonlinear device such as vacuum tubes, transistors, diodes, etc., the output of the device contains not two, but four (at least) frequencies. These are the two original frequencies plus the *sum* and *difference* frequencies.

For example, if two signals at 1.2 MHz and 1.655 MHz are heterodyned in a nonlinear device, the output will exhibit not only the original frequencies but also 1.2 + 1.655 MHz = 2.855 MHz and 1.655 − 1.2 MHz = 0.455 MHz. If either of the two original frequencies contain AM or FM modulation (voice or music, for example), the sum and difference frequencies will also contain this modulation.

In a superheterodyne receiver, one of the original signals is that which is transmitted by the radio station. This is frequently amplified prior to heterodyning, or mixing. The second signal is developed in the receiver itself by the local oscillator, the frequency of which is determined by the selector dial setting.

The tuning, or selector, dial is mechanically ganged to two variable capacitors, one for station tuning and the other for tuning the local oscillator. Thus, as the tuning dial is rotated to change stations, the local oscillator frequency changes proportionally. As the tuning dial is changed by the equivalent of 30 KHz, the frequency of the local oscillator changes by the same amount and in the same direction. For this reason, the difference frequency out of the mixer is always constant, typically 455 KHz, no matter what station the receiver is tuned for.

Because of the constant-frequency difference, the amplifier stages following the mixer, known as the *intermediate-frequency* (IF) stages, can be tuned for a single frequency. This results in far more efficient circuitry than having variable-frequency amplifiers. Hence, one or more IF stages are employed, tuned to 455 KHz.

14-4 TRANSISTOR RADIO—RF SECTION

In the circuit of Fig. 14-10, Q_1 functions as an RF amplifier and as the oscillator-mixer. The true collector load is the primary of T_1, and this is part of a resonant circuit. L_1 is the loopstick antenna that serves both as the input resonant circuit and as the antenna. It is caused to resonate with the tuning capacitor, which is "ganged" with the varia-

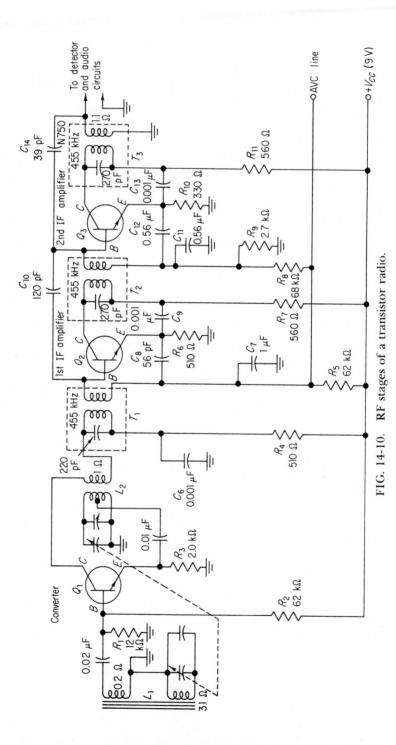

FIG. 14-10. **RF stages of a transistor radio.**

ble capacitor in the oscillator-tuned circuit L_2. L_2 provides the feed-back necessary to cause oscillations. The oscillator part of the circuit is a modified Armstrong oscillator (see Chap. 13), with one winding of L_2 in the collector circuit of Q_1. The local oscillator output is coupled to Q_1 through C_5 to the emitter of Q_1. Thus, by the principles of heterodyning, there will be four frequencies evident at the output of Q_1. These are the carrier of the station being received, the local oscillator output, and the sum and difference frequencies. The primary of T_1 is tuned to a fixed 455 kHz, and no matter where the tuning dial is set, the difference frequency will always be 455 kHz.

The RF amplifier-mixer Q_1 is operating in the familiar universal circuit. R_1 and R_2 form the base-biasing voltage divider, and R_3 is the emitter resistor. Because the dc resistance of the primary of T_1 is very small, we must modify the normal ERCA equation for this configuration to determine the quiescent condition of the circuit. If we consider that this dc resistance is zero, which is essentially true, we can easily calculate the measurable dc values in the circuit. As usual, the sum of RL and R_{CB} ratioed to RB must equal the ratio of RE to RB'.

$$\frac{RL + R_{CB}}{RB} = \frac{RE}{RB'}$$

Cross-multiplying,

$$RB \times RE = RB' \times (RL + R_{CB})$$

But RL is equal to zero in this circuit; so we can simply drop it from the equation.

$$RB \times RE = RB' \times R_{CB}$$

Solving for R_{CB},

$$R_{CB} = \frac{RB \times RE}{RB'} = \frac{62 \text{ k}\Omega \times 2 \text{ k}\Omega}{12 \text{ k}\Omega} = 10.3 \text{ k}\Omega$$

The gamma for this circuit is the ratio of R_{CB} to the total resistance (dc) in the collector and emitter circuit.

$$\gamma = \frac{R_{CB}}{R_T} = \frac{10.3 \text{ k}\Omega}{12.3 \text{ k}\Omega} = 0.84$$

This would normally indicate that the transistor is near cutoff, but since there is no actual dc resistance in the collector circuit, this is not quite the case.

To determine the base, emitter, and collector voltages, we must first determine collector current.

$$I_C = \frac{V_{CC}}{RT}$$

where RT is the sum of RE, R_{CB}, and R_4, since R_4 is in the path of total collector current.

$$I_C = \frac{V_{CC}}{RE + R_{CB} + R_4} = \frac{9}{2 \text{ k}\Omega + 10.3 \text{ k}\Omega + 0.51 \text{ k}\Omega} = 0.7 \text{ mA}$$

Now the drops around the emitter and collector circuits can be determined.

$$E_{RE} = I_E \times RE = 0.7 \text{ mA} \times 2 \text{ k}\Omega = 1.4 \text{ V}$$

$$E_{R_4} = I_C \times R_4 = 0.7 \text{ mA} \times 0.51 \text{ k}\Omega = 0.36 \text{ V}$$

$$V_C = V_{CC} - 0.36 = 8.6 \text{ V}$$

The base voltage V_B is about 1.5 V with respect to ground, because of the 0.1-V drop across the junctions. By using the simple relationships of earlier chapters we have determined the static condition of the mixer stage.

Q_2 is the first IF amplifier, and it is a fixed-frequency amplifier tuned to 455 kHz. The biasing of this stage is different from those encountered earlier. R_5 is part of this biasing circuit, and the automatic volume control (AVC) circuit actually determines the bias of Q_2. Without going into great detail, we can say that if the carrier of the station being received is very strong, the voltage from the AVC line is very negative. This will, of course, turn off somewhat the NPN transistor and will tend to keep the volume of this station at a low level. However, if the radio is tuned to a different and weaker station, the AVC line becomes more positive and Q_2 is turned on harder. The volume of this weaker station is therefore kept at about the same level as that of the stronger station. Since the bias of this stage of amplification depends upon the strength of the station being received, there is no simple way to calculate it.

To prevent amplified energy from Q_2 getting back to Q_1, which might cause unwanted feedback, R_7 and C_9 form a decoupling filter. Also, R_4 and C_6 in the collector circuit of Q_1 perform the same function. C_{10} and C_{14} are neutralizing capacitors to prevent oscillation in the IF amplifiers.

The second IF amplifier, Q_3, is very similar to the first, except that it is biased in the usual way. R_8 and R_9 form the base-circuit-biasing network, and R_{10} is the emitter resistor. R_{11} and C_{13} are the decoupling components to keep unwanted power-supply variations out of the stage.

The output of this stage of amplification is fed to the detector, usually a small-signal diode. The net result of this is to produce the audio-frequency signal that will be amplified by the audio amplifiers.

One further point should be stressed concerning this kind of circuit. We mentioned earlier that because there was no dc resistance in the collector circuit, the biasing of this circuit was somewhat different. When transformer coupling is used, the concepts of the collector and emitter voltage limits must be modified. In this case, the dc condition of the collector is that it is always at V_{CC}, regardless of the biasing condition. That is, the base can be set to any value between ground and V_{CC}, and the transistor will not saturate. However, as the base is biased on further, more and more collector current will flow, and it is very easy to damage a transistor by overbiasing it. For this reason, the transistor is usually biased so that the base voltage is no more than about 20 percent of V_{CC}. In the circuit shown for Q_1, 20 percent of 9 V would be 1.8 V, and so the base should not be more positive than this (NPN case). Since it is at 1.5 V, this is well within the limit.

14-5 COMPLEMENTARY TOTEM-POLE AMPLIFIER

The circuit illustrated in Fig. 14-11 is a special kind of amplifier, using two transistors of opposite kind but similar characteristics otherwise. It requires a single-ended input and provides a single-ended output. It is, in many respects, similar to a conventional emitter-follower. The circuit arrangement is taken directly from a commercial drawing, and it is difficult to analyze. Figure 14-12 illustrates the identical circuit, but is redrawn to show the true function of each component much more clearly.

Q_1 and Q_2 are *complementary* transistors; that is, one is NPN while the other is PNP. However, each is chosen from the same family, so that most major characteristics, such as h_{fe}, are similar. Because the bases are connected, each transistor will be affected in reverse to the

other by a given input signal. That is, a negative-going signal will
turn the NPN unit more off, but the PNP unit more on.

The circuit is used frequently for a power amplifier to deliver a
high-current signal to the load from a low-impedance source. The
circuit does not provide phase inversion from the output shown.

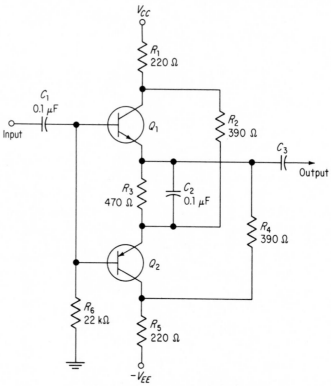

FIG. 14-11.   Complementary totem-pole amplifier.

However, an output may be taken from either collector that is out
of phase with the input. Note that *both* collectors are in phase with
each other. For example, if Q_1 is turned off somewhat by the input
signal, its collector will rise in the positive direction, while the same
input signal will turn Q_2 further on, also driving its collector *away*
from the -6-V source, or more positive.

Should it be necessary to dc-couple the output to the load, Fig.
14-13 shows a useful variation. Replacing R_3 with a 500-Ω potentiom-

eter will not only allow dc coupling, but will also provide for setting the output lead at exact ground potential to eliminate direct currents.

The reason for the five-resistor voltage divider is simply to prevent crossover distortion. Each transistor is biased on somewhat, ensuring that either one or the other is always turned on beyond the

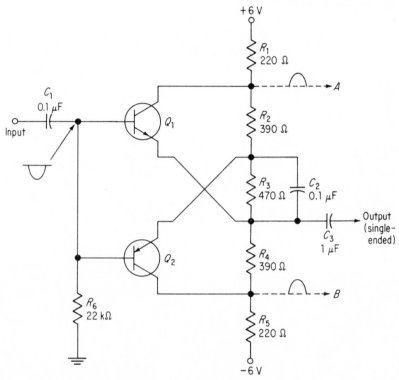

FIG. 14-12. The complementary totem-pole amplifier redrawn for clarity.

point where distortion occurs (in the very low-current part of the characteristics). Because the circuit is essentially an emitter-follower, no voltage gain is produced, but the current gain is approximately equal to the h_{fe} of the transistors.

14-6 CONSTANT-CURRENT SOURCES

A constant-current source is a device (or circuit configuration) that, under specified changing load conditions, will allow a current to flow

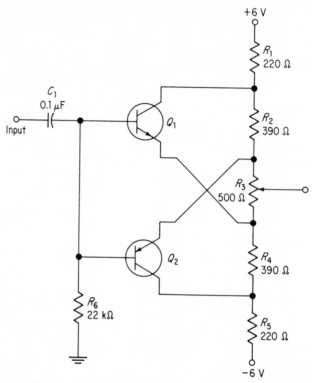

FIG. 14-13. The complementary totem-pole amplifier with provision for balanced dc output.

that is relatively constant. A very basic and simple constant-current source is illustrated in Fig. 14-14.

In this circuit, the source consists of a 100-V source with an internal impedance of 1 MΩ. With the switch in position 1, the short-circuit current, I_{sc}, is:

$$I_{sc} = \frac{E_{total}}{R_{total}} = \frac{1 \times 10^2}{1 \times 10^6} = 1 \times 10^{-4} = 100 \ \mu A$$

Now, changing the load resistance within limits will not substantially alter the total circuit current. Changing the switch to position 2 now alters the current as follows:

$$I_{total} = \frac{E_{total}}{R_{total}} = \frac{100}{1,000,010} = 99.999 \ \mu A$$

Note that the current is reduced by only 0.001 μA. For all practical purposes, the current is still 100 μA.

If the load resistance is changed to 100 Ω (position 4), the total circuit current is now

$$I_{\text{total}} = \frac{E_{\text{total}}}{R_{\text{total}}} = \frac{100}{1,000,100} = 99.99 \ \mu\text{A}$$

Current is now changed by only 0.01 μA and is still very close to the short-circuit value of 100 μA.

A bipolar transistor makes an excellent constant-current source, the reasons for which are illustrated in Fig. 14-15. Note that for large changes in V_{CE} the collector current changes very little. For this reason, bipolar transistors are often used as a constant-current source. Figure 14-16 illustrates such an instance.

Q_1 is biased in the center of its characteristics, and base voltage is held constant by R_1 and R_2. Constant base voltage results in constant base current, which in turn means constant collector current. As long as Q_1 is kept in the active region, one can vary the load resistance and the current through R_v will not change appreciably. This principle can easily be verified in the laboratory by inserting a milliammeter in series with R_v, or by measuring E_{R3}.

In commercial circuits, the base of Q_1 is often held more firmly constant by adding a large capacitor (even more effective is a zener diode) from base to ground. Further, R_3 is often bypassed with a large capacitor.

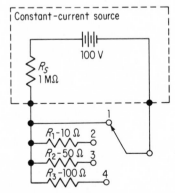

FIG. 14-14. Simple constant-current source example.

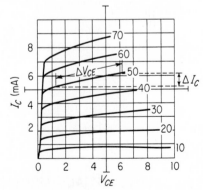

FIG. 14-15. Collector curves of a typical bipolar transistor illustrating the inherent constant-current characteristics.

In circuits such as this, the effective impedance of Q_1 is often several megohms, and yet it will function perfectly on small voltages. To illustrate this point, if Q_1 were to be replaced with a single resistor, to simulate a constant-current source, the supply voltage would necessarily be excessively large. For example, if the load current is 5 mA

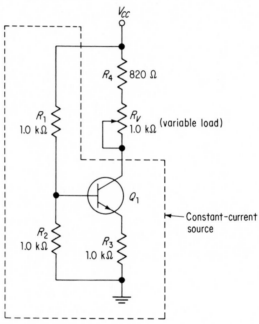

FIG. 14-16.   Bipolar transistor used as a constant-current source.

and Q_1 has an impedance of only 1 MΩ, the supply voltage for just the constant-current source would be

$$E = I \times R = 0.005 \times 1,000,000 = 5000 \text{ V}$$

The circuit of Fig. 14-16, on the other hand, will function with a supply voltage of 6 V, if a 3-V variation across the load resistor is acceptable.

14-7   DIFFERENTIAL AMPLIFIERS

In Chap. 12, a differential amplifier was discussed primarily in terms of analytical procedures. We shall now investigate a similar circuit with more attention given to circuit applications and circuit refinements. A differential amplifier is one in which the output is proportional to the instantaneous difference between two input signals. One application of the circuit uses the basic circuit properties to eliminate certain interference attendant upon long transmission lines. When long wires are used to transmit signals, they have a tendency to pick up interference, such as induced 60-Hz signals.

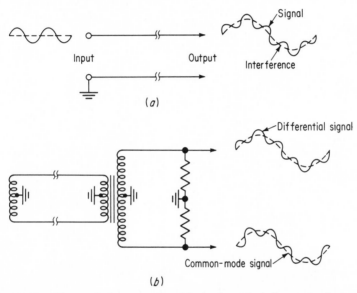

FIG. 14-17. Transmission systems: (a) single-ended, long-wired transmission system; (b) balanced (push-pull) long-wire transmission system.

In a single-ended transmission system, illustrated in Fig. 14-17a, the interference can seriously distort the desired signal. If the wanted and unwanted signals are of similar frequency, they may be nearly impossible to separate. If the two signals differ by several orders of magnitude, the interference can often be eliminated by suitable filters, but this is often not the case.

Figure 14-17b shows a balanced, or push-pull, system that allows the interference to be discriminated against, and in some cases to be virtually eliminated. The figure graphically illustrates the difference between the desired differential-mode signal and the undesired common-mode signal.

The balanced transmission line consists of parallel wires that are *not* at ground potential, plus one ground conductor. Therefore, if both signal wires are very close to each other over the length of the transmission path, each will pick up the *same* interference signal. This interference signal is termed a *common-mode* signal, since it is common to both of the signal wires. In both polarity and amplitude the common-mode signal is the same on both wires. The differential signal, however, has the same amplitude, but is of *opposite polarity* on each wire. The differential amplifier, then, is specifically designed to minimize the common-mode signal, while maximizing the differential signal.

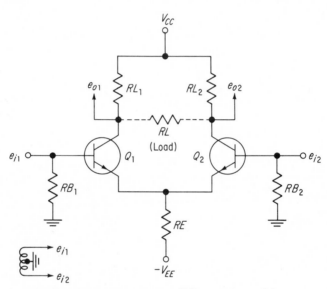

FIG. 14-18.   Circuit of a differential amplifier.

A typical differential amplifier is shown in Fig. 14-18. The input signal originates at a transducer of any sort that will produce a differential, or push-pull, signal. Each of the input signals is presented to its respective base to be amplified. The outputs are taken from each collector and delivered to the load, as indicated. In an ideal circuit, the common-mode signals will be greatly attentuated across the load, while the differential-mode signals will be amplified.

The amplifier itself is a balanced, or symmetrical, circuit, in that it has two inputs and two outputs, and each half of the circuit appears identical to the other. Indeed, the more nearly identical the two halves can be made to be, the better the common-mode rejection.

Basically, the circuit action is very simple. If the base of Q_1 is driven more positive than its quiescent value, the voltage at this collector will become more negative, and will of course be amplified somewhat. At the same time, the base of Q_2 will be driven more negative, and its collector will go more positive. This, of course, is conventional push-pull action. If Q_1 and Q_2 have similar characteristics, and if RL_1 and RL_2 are the same value, both outputs will have the same amplitude but opposite polarity. Each collector voltage will be 180° out of phase with its respective base voltage.

In a perfectly symmetrical circuit, the voltage at each collector will be the same, with no signal applied. There will therefore be no current in the load with no applied signal. However, with an applied differential signal, the collector voltages will vary around the quiescent value, and signal current will flow in the load. Thus, an amplified replica of the differential signal will appear across the load.

Now, assume that a common-mode signal appears at the input to the circuit. In this instance, both bases are driven in the same direction by the same amount, and so the collectors are also driven in the same direction and by the same amount. The voltage at each collector will change, relative to ground, by the same amount. Therefore the voltage across the load will *remain the same*. It is clear that no current flows in the load for common-mode inputs, and the circuit has effectively reduced or eliminated them.

As an example, assume the quiescent collector-to-collector voltage is zero, with e_{o1} equal to +5 V relative to ground. Also, e_{o2} is equal to +5 V relative to ground. The current through the load resistor is therefore zero. If now the common-mode signal causes the collectors to go to +4 V, relative to ground, the current through the load resistor is still zero! The circuit has for all practical purposes not produced an output for this common-mode input.

Basically, then, the circuit functions as described. In practice, the circuit does not reject the common-mode signal completely, since to do this, each half of the circuit must be *exactly* the same as the other, and this is not possible for ordinary production-run equipment. However, in the design of such a circuit many things can be done to make the common-mode rejection as high as possible. One of these is to provide a constant-current source for the emitter circuit; i.e., make RE as high in value as possible. Better yet, place a transistor, configured as a constant-current source, in the emitter circuit. Such a modification is illustrated in Fig. 14-19. Other methods that will yield higher rejection ratios are handpicking the components for similarity and introducing negative feedback for the common-mode signal. One of the big advantages of integrated circuits is that all components of the amplifier are constructed on a single chip of silicon, and this greatly improves many of the features of the differential amplifier. This will be discussed in more detail in Chap. 17.

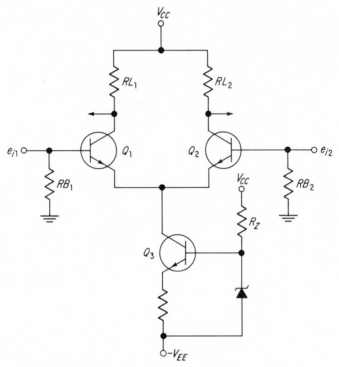

FIG. 14-19.   Differential amplifier with constant-current source.

14-8   OPERATIONAL AMPLIFIERS

As usually implemented, the operational amplifier is simply a differential amplifier with certain external circuit additions that significantly alter the basic operation. The circuit is usually defined as one that has very high voltage gain, is direct-coupled, and uses external feedback to control the response characteristics. The ideal operational amplifier is one that is dc-coupled, has an infinite voltage gain, has an infinite input impedance, and has an output impedance equal to zero. In practice, such stringent requirements cannot be attained, but can be approached very closely by good circuit design.

The major desirable characteristic of this circuit is that it can be designed to have extremely good long-term gain stability. That is, the voltage gain of a simple amplifier will be found to vary as the equipment ages. This is due to many factors, which include changes in transistor characteristics, circuit component changes, temperature

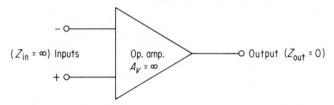

FIG. 14-20.   Diagram of ideal operational amplifier.

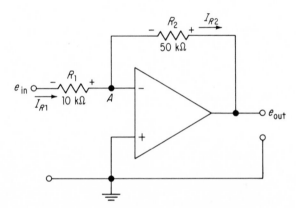

FIG. 14-21.   Operational amplifier with external negative feedback.

variations, humidity variations, etc. All or any of these can, and will, cause the voltage gain of many circuits to change. The operational amplifier is specifically designed to minimize these effects.

The name of this circuit derives from the fact that its original application was in analog computers, and it was used to perform mathematical "operations" such as addition, subtraction, etc. It has found widespread use in many other fields in recent years, but the name persists. Figures 14-20, 14-21, and 14-22 will serve to illustrate the basic principles of the operational amplifier. The first of these shows the ideal building block requirements for the circuit. Shown are the high input impedance and low output impedance, plus the very high voltage gain and two inputs. The two inputs are labeled plus and minus; the positive input simply indicates that the output will *not* be inverted, while the negative input *will* be inverted. The second draw-ing shows the complete circuit, with the external feedback clearly indicated.

Figure 14-22 illustrates a practical circuit that we shall investi-gate. It must first be understood that the symbol for the amplifier

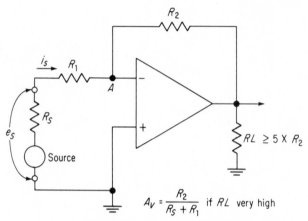

FIG. 14-22.   Realistic representation of the typical operational amplifier.

itself, the triangle, represents a very high gain amplifier with at-tendent high input resistance. The actual circuit for this configuration will be given following this general discussion. The signal source, as does any source, has a certain value of internal resistance. This is represented by R_S in the figure. By design, R_S is made to be as small as possible. If it can be made to be insignificant relative to R_1, then it can be ignored in any calculations. R_2 is the feedback resistor. Its purpose is to provide a path for a portion of the output signal to be applied back to the input to be summed with the input signal. That is, at the junction of R_1 and R_2, there are two currents: one from the source (signal) and one from the output of the amplifier itself. Pro-viding that the input resistance of the amplifier is very high, essen-tially no current flows into the amplifier input terminal. For this rea-son, the currents flowing in R_1 and R_2 are a function only of the values of the resistors, and of course of the voltage across them. These two resistors, then, will be found to be the only determining factor in setting the voltage gain of the amplifier. Hence, if the resistors them-selves are stable over long periods of time, the gain of the amplifier will be just as stable.

A simple example will be used to describe the circuit action of this operational amplifier. Assume first that the voltage gain of the amplifier with no feedback is at least 10,000. Furthermore, assume that R_2 is equal to 100,000 Ω while R_1 is 1000 Ω. For simplicity, the value of R_S will be assumed to be essentially zero. The signal source is generating a signal e_s that appears across its terminals. If e_s is 10 mV peak to peak, this voltage will cause some value of current to flow into R_1. To determine the value of this current, the voltage across R_1 must be found. As will later be proved, the junction of R_1 and R_2 is a "virtual ground." That is, while it is not physically connected to ground, it is always at, for all practical purposes, ground potential. Since this is true, the value of current through R_1 can now be found.

$$i_s = I_{R_1} = \frac{e_s}{R_1} = \frac{0.01}{1000} = 10 \ \mu A \text{ peak to peak}$$

Now, because for all practical purposes the input resistance of the amplifier is infinitely high, the 10-μA input current can only flow through R_2, the feedback resistor. The voltage drop across R_2 is therefore determined by this current.

$$E_{R_2} = i_s \times R_2 = (10 \times 10^{-6})(1 \times 10^5) = 1 \ V$$

Again, because the junction of R_1 and R_2 is a virtual ground, and because the output is measured in reference to ground, the voltage across R_2 is the output. It can now be appreciated that the voltage across R_1 is the input voltage, while the voltage across R_2 is the output voltage. The voltage gain can now be found.

$$A_v = \frac{e_o}{e_i} = \frac{i_s R_2}{i_s R_1} = \frac{R_2}{R_1} = \frac{1 \times 10^5}{1 \times 10^3} = 100$$

Thus, the circuit yields a voltage gain that is a function only of the two resistors, rather than any characteristics of the amplifier itself. Other factors, such as input-output impedances and frequency response, are also held to equally tight tolerances.

To verify that the junction of R_1 and R_2 is a virtual ground, it must be realized that R_1 and R_2 form a voltage divider. Because the overall voltage gain is 100, the output voltage will always be 100 times the input voltage. First, assume the source voltage to be 1 mV; the output voltage is then 100 mV. Now, the value of current through the divider can be found.

$$i_s = \frac{e_s + e_o}{R_1 + R_2} = \frac{0.101}{101,000} = 1 \ \mu A$$

The voltage drops across the individual resistors will be found to be 0.001 V for R_1 and 0.1 V for R_2. Equating the circuit voltage drops for the point in the circuit in question (point A)

$$e_A = e_s - i_s R_1 = 0.001 - 0.001 = 0 \ V$$

For this value of input voltage, the point is truly at ground, or zero, potential. To verify that this is always the case, simply assign a different value of input, for example, 10 mV.

$$i_s = 10 \ \mu A$$

The voltage drop across R_1 is now 10 mV, and the drop across R_2 is 1.0 V. Again, the voltage at point A in reference to ground is

$$e_A = e_s - i_s R_1 = 0.01 - 0.01 = 0 \ V$$

Hence, with any value of input, point A is always at ground potential.

A circuit diagram for an operational amplifier is shown in Fig. 14-23. It consists of two cascaded differential amplifiers. Clearly shown are the two inputs, and the connection points for the two added resistors R_1 and R_2. If the individual stage gain for each differential amplifier is 100, then the overall gain with no feedback is simply the product of the two gains, or 10,000. In Chap. 17 an operational amplifier will be shown that uses integrated-circuit techniques.

14-9 COMPLEMENTARY FEEDBACK AMPLIFIER

The complementary feedback amplifier, illustrated in Fig. 14-24, is used frequently in applications where a definite low- or high-frequency cutoff is required. In such a case, a reactive load is placed across the terminals labeled A, and this alters the frequency response. A capacitor across these terminals results in a high-boost effect, while an inductor yields a low-boost effect.

Basically, the circuit functions as follows. Q_1 is an NPN unit, while Q_2 is PNP. Each is acting as a common-emitter amplifier, with the output of each fed back to the input of the other. That is, the collector of Q_1 is connected directly to the base of Q_2, while the collector of Q_2 is tied to the emitter of Q_1 via R_2. The circuit is heavily degenerative.

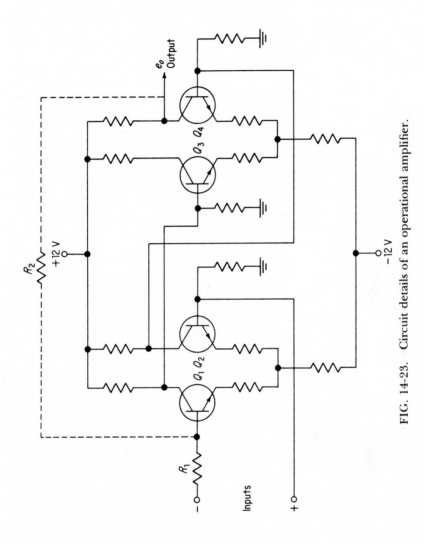

FIG. 14-23.   Circuit details of an operational amplifier.

With normal values of resistances typically used, the voltage gain is relatively low, on the order of 10 or so.

Tracing a signal through the amplifier will show how each transistor interacts with the other. Assume a small positive-going voltage at the base of Q_1. This is amplified and inverted by Q_1 and presented to the base of Q_2. A negative-going voltage on the base of Q_2 is amplified and inverted and appears as a positive-going change at the junc-

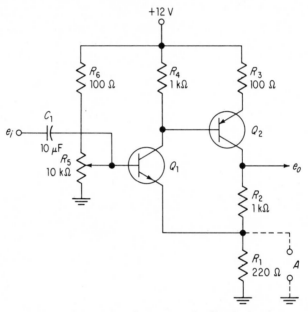

FIG. 14-24.   Complementary feedback circuit.

tion of R_1 and R_2, which is the emitter of Q_1. Now, a positive change at the emitter of Q_1 will tend to cancel the effect of the original positive voltage at the base. Hence this is negative feedback.

The voltage gain of this amplifier is determined by the two resistors R_1 and R_2. This relationship is given below using the values given in the circuit diagram.

$$A_v = 1 + \frac{R_2}{R_1} = 1 + \frac{1 \text{ k}\Omega}{220} = 5.55$$

Circuit variations can cause the overall response to be changed in a wide variety of ways. Figure 14-25 illustrates four variations. By bypassing R_1, the high frequencies are boosted by virtue of the fact that as the reactance of C becomes less, the feedback becomes less and less, allowing the gain to increase. If R_2 is bypassed, the circuit yields more gain at the low frequencies, and becomes simply an emitter-follower if X_C is essentially zero. If a series-tuned circuit is added across R_1, the feedback will be reduced at the resonant fre-

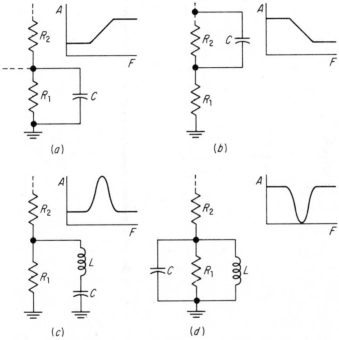

FIG. 14-25. Frequency-response variations: (a) high-boost; (b) low-boost; (c) bandpass; (d) band-reject.

quency of L and C. This allows greater amplification at this frequency. The bandpass will be determined by the Q of the circuit. Finally, if a parallel-tuned circuit is added across R_1, the circuit will perform as a band-reject circuit.

The circuit is difficult to bias properly. Hence the potentiometer allows proper bias to be determined empirically. If the circuit is to be used as a laboratory experiment, the circuit can be adjusted for proper bias by applying a signal and varying the potentiometer such that the output, viewed at the collector of Q_2, does not clip either the negative or the positive peaks. This requires a continuously variable input amplitude. Obviously, both transistors must be operated as near the center of their characteristics as possible.

14-10 THE DARLINGTON CONNECTION

The Darlington (or superbeta) connection is illustrated in Fig. 14-26a. Because the beta of a typical transistor is in the neighborhood of 50

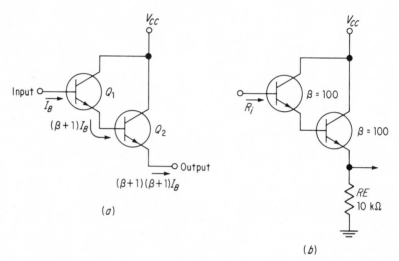

FIG. 14-26. (a) Darlington connection. (b) Illustrating the high input resistance.

to 200, this circuit is used to simulate a transistor whose beta is very much higher than this. Two separate transistors are connected as shown, with the emitter of the first unit driving the base of the second, and with both collectors connected. Often manufacturers provide just such an arrangement in a single package, consisting of either two transistors connected together or a similar arrangement constructed upon integrated-circuit principles.

To understand the circuit function, assume that a small current is flowing in the base lead of Q_1. The emitter current of this transistor is therefore $(\beta + 1)$ times greater. Now, this amount of current becomes the base current for Q_2, whose emitter current is $(\beta + 1)$ times greater than its base current. Thus, the output current is $(\beta+1)(\beta+1)$

times the input base current. If both transistors have about equal betas, this can be simplified to read:

$$I_{out} \cong \beta^2 I_B$$

One of the most useful applications of such a configuration is to provide a circuit with extremely high input resistance. Such a circuit is illustrated in Fig. 14-26b, where the input resistance is approximately $\beta^2 RE$.

$$R_i = 100^2 \times 10,000 = 1 \times 10^8 \ \Omega$$

This is a very high input resistance for transistor circuits. One word of caution regarding this circuit type: even though the beta is very desirably higher, this attribute cannot be had for nothing. Because of the greatly increased current gain, the temperature response is degraded by the same amount as the current gain. Hence, great care must be exercised in the design of such a circuit to ensure that a temperature increase will not interfere with proper operation.

14-11 TRANSFORMERLESS POWER AMPLIFIER

Using iron-core transformers makes it difficult to obtain good frequency response at both high and low frequencies. Good low-frequency reproduction requires a transformer with a large amount of iron, while the very nature of the inductance itself makes for poor high-frequency response. High-quality audio transformers are very expensive, and the advent of transistors has given the design engineer an alternate choice. A typical transformerless power amplifier is shown in Fig. 14-27.

The circuit shown is a direct-coupled power amplifier with good low-frequency response. The two output transistors, Q_4 and Q_5, operate in a class B single-ended push-pull arrangement. The drivers and phase inverters, Q_2 and Q_3, also operate class B, and appear Darlington-connected to the output stage. Because Q_3 is a PNP transistor, it provides the necessary phase inversion. Q_4 and Q_5 are operated with a small forward bias to eliminate crossover distortion. This bias is set by the voltage drop across the 470-Ω resistors. Q_2 and Q_3 are also operated with a small forward voltage that derives from the two silicon diodes D_1 and D_2. Q_1 is a preamplifier and driver that is operated class A. The 22-Ω resistor in series with the 0.22-μF capacitor prevents feedback at the higher frequencies where the possibility of phase reversal could cause oscillation. The circuit as shown will provide approximately 3 W of power into the load with an input signal of 350 mV.

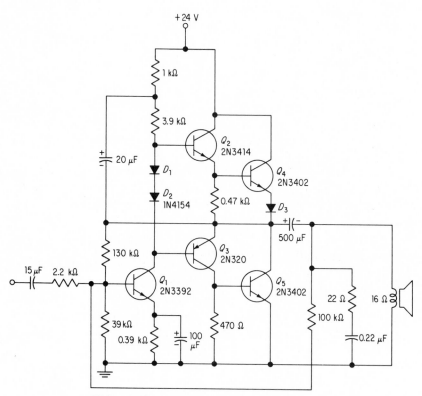

FIG. 14-27.   Transformerless output stage.

QUESTIONS AND PROBLEMS

Select the correct answer to each of Questions 14-1 to 14-6.

14-1   A class A amplifier is one in which
  (*a*) I_C flows most of the time
  (*b*) the base is biased just to the cutoff point
  (*c*) emitter current flows all the time
  (*d*) the collector voltage V_C often rises to V_{CC}

14-2 A class A amplifier is used when
(a) the highest gain is desired
(b) no phase inversion is wanted
(c) minimum distortion is to be achieved
(d) unidirectional voltages are to be amplified

14-3 A class A amplifier is usually biased, for large signals,
(a) so that V_C can swing equal distances either side of the quiescent point
(b) just at cutoff
(c) far below cutoff
(d) at the point where $V_C = V_E$

14-4 A class B amplifier is biased
(a) about 2 times cutoff
(b) just at cutoff
(c) at the midpoint of the load line
(d) so that I_B just equals I_C

14-5 A class B single-ended amplifier is useful because
(a) the input signal is amplified linearly
(b) the gain is very low
(c) the output is an exact replica of the input
(d) I_C flows for 180° of the input cycle

14-6 When a transistor is operating with $I_C = V_{CC}/RL$, it is said to be
(a) in the active region
(b) in cutoff
(c) in saturation
(d) amplifying linearly
(e) a class B amplifier

14-7 Refer to Fig. 14-10. Discuss the bias arrangement of Q_2.

14-8 Refer to Fig. 14-10. In the base circuit of Q_1, R_2 is to be changed to a 47-kΩ resistor. What is the new value of V_B? V_E?

14-9 Refer to Fig. 14-10. Briefly discuss the purpose of L_2.

14-10 Refer to Fig. 14-10. What component(s) constitute(s) the collector load for Q_2?

CHAPTER 15 POWER SUPPLIES

The power supply is a most important part of the overall equipment. Since an entire piece of equipment, whether a computer or a TV set, depends upon the power supply to deliver the proper voltage and current, failure of the supply can, and usually does, mean failure of the entire equipment. A power supply can fail in many ways, not all of which are obvious. Trouble in a piece of electronic equipment is often due to faulty power-supply operation.

A typical power supply is shown in Fig. 15-1 in block-diagram form. Its main purpose, of course, is to provide the electronic circuits with the proper value of dc voltage and current from the ac supply lines. The power transformer serves two purposes, one of

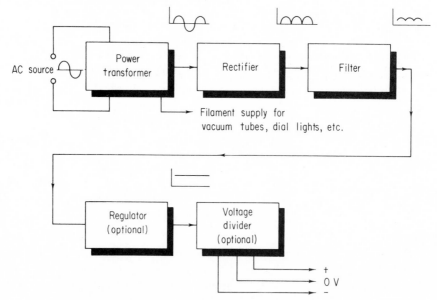

FIG. 15-1.   Block diagram of typical power supply.

them being to isolate the equipment from the power lines (ac) to eliminate the shock hazard. Since one side of the ac line is grounded to earth, a definite shock hazard exists, which the transformer eliminates, there being no physical connection between the primary and secondary windings. The second function is to provide the proper voltages to the rest of the circuit to produce, ultimately, the correct dc and ac voltages by stepping up or down the ac line voltage.

The rectifiers serve to convert the ac from the transformer to a unidirectional voltage and current (dc) that is required by most electronic circuitry. The rectified ac is a pulsating dc and, since the value is constantly changing, cannot be used by itself. The filter circuit then smooths out the pulsations and helps to form much better dc, with only a small "ripple" voltage left. If a regulator is to be incorporated, its purpose is to ensure that, with normal load-current variations, the dc output voltage of the power supply will remain constant.

Finally, the voltage divider, if used, can provide various and sundry values of dc voltage less than the maximum output of the supply.

Of the many kinds of rectifier circuits possible, we shall be interested in but a few. These are the half-wave rectifier, the full-wave rectifier, the full-wave bridge, and the voltage doubler. The simplest of these is the half-wave rectifier. We shall consider this circuit first.

15-1 THE HALF-WAVE RECTIFIER

Because of a number of disadvantages the half-wave circuit is seldom used, except in cases where the advantage of the cost factor outweighs the drawbacks. A typical circuit is shown in Fig. 15-2, and a brief description is given below, along with a list of the various qualities of the general circuit.

The secondary of the transformer produces an ac voltage, 60 Hz,

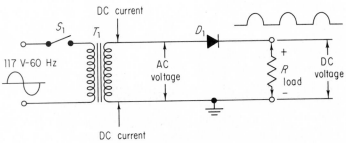

FIG. 15-2.   Halfwave-rectifier circuit, + voltage.

of a suitable value to provide the necessary dc voltage. The secondary voltage causes the anode of the diode to be at one instant more positive than its cathode, at which time the diode will conduct. On the next alternation, the secondary voltage is such as to make the anode more negative than the cathode, and the diode will not conduct. The current through the diode and the load is flowing in one direction, and so is a dc current, although not a smooth dc. Note that for an entire half-cycle, 8.3 ms (milliseconds), no current at all flows. So the delivery of power to the load is done quite inefficiently.

The rectifier itself can be either a vacuum diode or a solid-state diode. We shall confine our discussion to the solid-state devices.

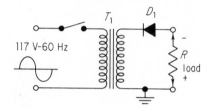

FIG. 15-3.   Halfwave-rectifier circuit, − voltage.

Diode D_1, it should be noted, is carrying the full load current, and must be rated for at least the maximum load current to be encountered. Also, the full peak voltage of the secondary winding appears across the diode when the polarity is such as to cause reverse bias. The diode must be able to withstand this without breaking down. The diode peak inverse voltage (PIV) must exceed the highest peak voltage from the secondary.

The circuit as shown in Fig. 15-2 would be called a *positive supply*, since the output is more positive than ground. Figure 15-3 shows the same circuit connected to produce a negative voltage. The only difference between the two circuits is the direction in which the diode is installed.

It is possible to extract both polarities at the same time from the same transformer. This is illustrated in Fig. 15-4.

When the diodes are presented a positive voltage with respect to ground, D_2 will conduct, and with a negative voltage, D_1 will conduct. Thus across each load is developed a voltage that is in one case positive to ground, and in the other case negative to ground.

A dc voltmeter reading taken across the load of Fig. 15-3 or 15-4 would yield a reading of 0.45 times the rms secondary voltage. This low figure is the major reason why the half-wave rectifier is seldom used.

The half-wave rectifier has few advantages, among which are the few parts required and the consequent low cost. Among the disadvantages are an overly large transformer, stringent filtering requirements, low-output-current drain, and poor regulation.

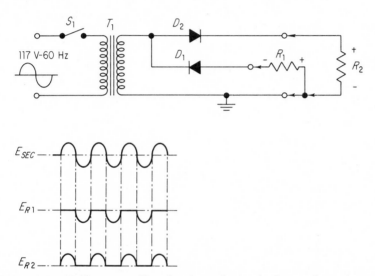

FIG. 15-4.   Halfwave-rectifier circuit producing both + and − voltages.

15-2 THE FULL-WAVE RECTIFIER

The full-wave circuit, shown in Fig. 15-5, is probably the most widely used of all rectifier circuits. Here the transformer is center-tapped, and the addition of the extra diode allows the development of voltage to the load during *both* half-cycles, thus providing twice the energy for the load. With the secondary polarity such as to make the upper end positive, D_1 is conducting and I_1 is flowing. This current must flow through the load and develops the polarity shown. The next half-cycle, the transformer polarity is the reverse, and D_2 now conducts. As a result, current again flows through the load in the same direction as before.

The net result is a much better dc developed across the load, which will require far less filtering to smooth out. But a price must be paid for the efficiency gained. If the transformer secondary yields

500 V rms from end to end, each diode in its turn is working with only half this much voltage. Thus the output is derived from a source that is producing 250 V, and a meter across the load resistor would actually read about 225 V. Thus, because only half of the secondary winding is used at a time, the output is about half of the total second-

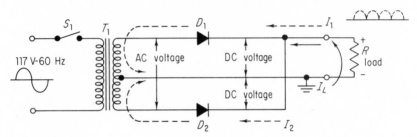

FIG. 15-5.   Fullwave-rectifier circuit.

ary voltage. However, the ripple frequency is double that of the half-wave rectifier, and so will be much easier to smooth out. A dc voltmeter placed across the load will read 0.9 times the rms value of one-half of the full secondary voltage.

Some of the advantages of a full-wave rectifier are simple filtering requirements, smaller transformer size than would be expected in view of the lack of dc in the secondary coil, high ripple frequencies, and good efficiency.

15-3   THE BRIDGE RECTIFIER

The bridge rectifier is in many respects similar to the full-wave rectifier just discussed. In terms of ripple frequency and efficiency, the two are quite similar, but in terms of output, they are quite dissimilar.

Consider the circuit of Fig. 15-6. Note that the bridge is across

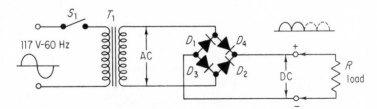

FIG. 15-6.   Fullwave bridge-rectifier circuit.

the full transformer secondary, and the voltage delivered to the load is derived from the full secondary voltage without a center tap. When the secondary voltage makes the upper end positive, current will flow through D_3 and D_4 in a direction to produce the indicated voltage drop across the load. When the secondary voltage reverses in polarity, D_1 and D_2 will conduct. Now the current will flow in the load in the same direction as before. A dc voltmeter placed across the load will read 0.9 times the full secondary rms voltage.

The reason for the factor of 0.9 as used above is rather interesting. The rms value of secondary voltage is the applied voltage. The dc meter used across the output will deflect in proportion to the average of the pulsating waveform. The rms value in terms of peak is 0.707, and the average value in terms of peak is 0.636. The dc meter will therefore read proportionally lower than the rms value by the ratio $0.636/0.707 = 0.9$, and if 100 V rms is applied, the dc output must be $(0.636/0.707) \times 100 = 90$ V, assuming no voltage drop in the secondary winding or across the diodes.

The bridge rectifier has the advantage of being able to use the entire secondary voltage and still provide full-wave rectification.

15-4   THE VOLTAGE DOUBLER

The half-wave voltage doubler shown in Fig. 15-7 is a rectifier circuit arranged to produce an output voltage up to twice the applied voltage. In fact, the output can be made to become greater than twice the rms input voltage.

To understand how the circuit works, consider D_1 and C_1 by themselves, as in Fig. 15-8. Here only with the polarity shown can D_1 conduct, and then only until C_1 becomes charged. C_1 will charge as the input builds up to peak. As the input falls away toward zero, the capacitor cannot discharge, for to do so would require that it discharge through D_1, anode to cathode; the diode will not allow this, of course. Thus the voltage across the capacitor remains at the peak value.

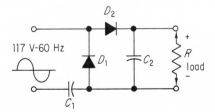

FIG. 15-7. Voltage-doubler circuit.

Now, when the input changes polarity, as shown in Fig. 15-9a, the discharge path of C_1 is as shown in Fig. 15-9b, where the circuit is redrawn.

Capacitor C_2 now sees *two voltage sources*, the ac supply *and* C_1. The circuit connection places these two sources in series aiding, and

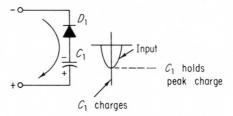

FIG. 15-8.   Illustrating the charging of C_1.

C_2 charges to twice the *peak* value of the applied voltage. As an example, if the circuit is operated off the ac line, the input is on the order of 117 V rms. The voltage across C_2, then, must be (117 × 1.414)(2) = 331 V. This would be true only if the load current were very small. As soon as the load current became appreciable, compared with the storage capacity of the capacitors, the voltage across C_2 would sag, since the capacitors could now discharge a large amount between cycles. For this reason, C_1 and C_2 are made large enough to produce the desired output with normal load. Values in the vicinity of 100 μF or greater are typical for C_1 and C_2.

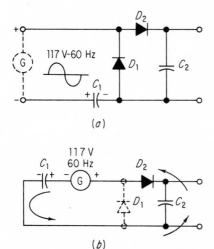

FIG. 15-9.   The power source and C_1 in series aiding charge C_2 to twice the peak voltage of the source.

15-5 THE CAPACITOR-INPUT FILTER

The need for a filter is easy to understand when one looks at the output voltage from a rectifier. Ideally, the output would be a smooth dc. But we find that the rectifier produces a pulsating dc that is anything but pure dc. There is still a large ac component, called *ripple,*

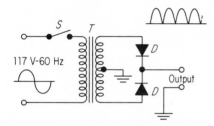

FIG. 15-10. Unfiltered fullwave-rectified output.

left, even though the current never changes direction. The purpose of the filter is to remove all, or part of, the ripple.

The unfiltered output of a full-wave rectifier is shown in Fig. 15-10. Note that the areas between peaks in the waveform are quite large and represent little or no energy delivered to the load. A way of filling in these areas is shown in Fig. 15-11. Here we have added a filter capacitor that will charge up to the peak value after several cycles and, connected in this way, will produce reasonably good dc.

If the power supply did not have to supply other circuits with dc power, the filter capacitor shown above would suffice. But as soon as the power supply begins to deliver current, the capacitor is called upon to provide current to the load. The more current that is drawn from the supply, the more current is taken from the capacitor as well as from the power supply. We know that a discharging capacitor suffers a voltage loss. Under these conditions, the capacitor can no longer hold up the dc level, and the output voltage drops. Because of this,

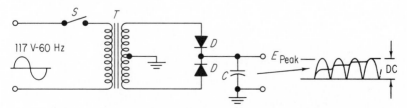

FIG. 15-11. Filtered fullwave-rectified output.

the ripple voltage becomes greater as more current is taken by the load. This idea is illustrated in Fig. 15-12.

The capacitor by itself will not filter the ripple as well as might be desired, and a very useful filter is shown in Fig. 15-13. It uses both inductance and capacitance to accomplish its job.

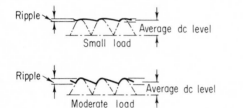

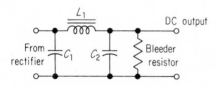

FIG. 15-12. The changing dc level and ripple with changing load current.

FIG. 15-13. Capacitor-input filter.

This circuit is called a *capacitor-input filter*, and will smooth the output of a full-wave rectifier very well. The inductor L_1 opposes any change in current through itself. Since the load current must flow through it, it tends to keep the current constant. With the additional capacitor, the output is virtually free of ripple. Typical values might be $C_1 = 200 \ \mu F$, $C_2 = 200 \ \mu F$, and L_1 in the range of 1 to 30 H (henrys).

The usefulness of the choke coil can be realized by simply calculating the inductor's reactance at the ripple frequency. If the rectifier is a full-wave type operating on 60 Hz, the ripple frequency must be 120 Hz. If the choke is 30 H, $X_L = 2\pi FL = 6.28(1.2 \times 10^2) \times 30 = 22,600 \ \Omega$.

Since the dc resistance is usually very low in value, the dc drop is very small, but the reactance is very large, and the choke opposes to a large degree any variation, or ripple, left by C_1.

The capacitor-input filter has the following characteristics:

1.   High voltage output
2.   Poor regulation

3. High peak currents in the rectifiers
4. Small maximum current delivered to the load, compared with certain other types

The high voltage output can be explained by the fact that the input capacitor tends to charge to the peak value of the rectified input. However, if the load current increases, the capacitor discharges more deeply between the peaks, and the average voltage drops. A changing dc voltage with changing load current signifies poor regulation. Where good regulation is required, this circuit is not used.

The high peak currents occur because of the need to charge C_1 through no limiting resistance. The larger the load current and the larger C_1, the greater the initial charging current.

The small output current exists because the maximum power that can be dissipated across the load is determined by the size of the components, notably the transformer. If, for instance, a transformer can deliver 100 W across the load resistor, 100 W could be produced by

400 V at 0.25 A   50 V at 2 A
200 V at 0.5 A   25 V at 4 A
100 V at 1 A

That is, the higher the output voltage, the lower the allowable maximum current to keep $E \times I$ at, or below, 100 W.

Closely associated with the filter is the *bleeder* resistor, and this serves to discharge the capacitors when power is removed to avoid the possibility of shock hazard. Also, it serves to limit the no-load current to a value greater than zero, and will improve the regulation of the supply. Bleeder current is often about 10 percent of the total load current.

15-6 THE CHOKE-INPUT FILTER

Another kind of filter is the *choke-input filter*, shown schematically in Fig. 15-14.

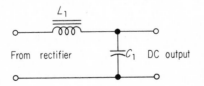

FIG. 15-14. Choke-input filter.

For given-size components this filter reduces the ripple somewhat less than the capacitor-input filter because of the smaller amount of capacitance. However, it possesses other virtues that make it very useful in certain circumstances. For instance, it has far better regulation, and is generally used in cases where the load current changes quite drastically. The dc output voltage tends to remain more stable in this circuit, compared with the capacitor-input filter. Another advantage is that the maximum dc current can be much higher, since the voltage is somewhat less than an equivalent capacitor-input circuit. Since the choke has a large reactance, the peak current is limited to a smaller value, and usually smaller rectifiers can be used.

15-7 ELECTRONIC REGULATORS

For supplying certain types of circuitry with dc power, a conventional filtered supply is inadequate. Thus it becomes necessary to improve the regulation of the power supply by electronic means. Essentially, electronic regulators can be divided into four types.

1. The breakdown diode, or zener diode, regulated supply, constant voltage output
2. The shunt regulator, constant voltage output
3. The series regulator, constant voltage output
4. The series regulator, constant current output

The discussions to follow will be confined to the first three types.

The Zener Regulator

The simplest of the regulator circuits is the zener-diode-regulated power supply, shown in part in Fig. 15-15.

Since the zener diode has, within certain current limits, a constant voltage drop across it, the voltage delivered to the load is equally constant. If it is considered that the zener diode is a kind of variable resistor, it will be easy to understand how the circuit works.

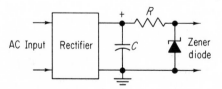

FIG. 15-15.   Zener diode regulation.

Figure 15-16 shows an equivalent circuit where R_z is the resistance of the zener. Suppose the load resistance decreases, and so draws more current. The series resistor R would cause the output voltage to decrease by the additional drop across R. However, as the voltage begins to sag by only a few millivolts, the internal resistance of the

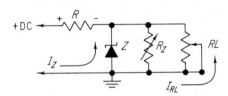

FIG. 15-16.   Zener resistance changing with varying load.

zener diode becomes larger, and so passes less current. Since I_z must also flow through R, the drop across R will be *less* now, since less current than before is flowing. The output voltage tends to remain constant, even though the load current varies. If the load current decreases, the drop across R is less, and the voltage across RL tends to become greater. The resistance of the zener now decreases, drawing *more* current than before, thus causing a greater drop across R. Again the load voltage tends to remain constant.

Within rather narrow current limits, the zener diode will maintain the output voltage within a few hundred millivolts of rated value. That is, a typical zener, with internal current from about 5 to 25 mA, will allow the voltage to vary from 6.3 V at 5 mA to 6.55 V at 25 mA. For many applications this can be considered a constant voltage, since the difference between the two values is only 0.25 V.

Shunt Regulation

The second type listed above is the shunt regulator, which is quite similar to the zener regulator just discussed. The circuit is shown in Fig. 15-17, and the similarity is readily seen. The only significant

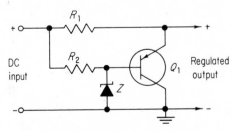

FIG. 15-17.   Shunt regulator.

difference is that the regulating current is about beta times greater than with the zener alone. Note that Q_1 is really an emitter-follower (common collector), and its emitter load resistance is the load itself, which may in turn consist of several transistor circuits.

Series Regulation

The third kind of regulator is the series type, Fig. 15-18, and here the regulating element is in series with the load current.

In this circuit, Q_2 is the series regulator, and again can be considered a simple variable resistor, except that in this case it is in series with the load. As in all regulating systems, the unregulated input must be somewhat greater in amplitude than the regulated output. In this case it is about 11 V greater. Q_2 is an emitter-follower, and its load is the electronic circuitry being operated from this supply. This being true, the regulator is operating in the active region, and will offer quite a bit of resistance to the circuit. R_3 is a potentiometer that adjusts the output to exactly the required value, and is part of a voltage divider which has a constant 6.3 V across it due to the zener diode.

Assuming the circuit is performing properly, suppose the demand for load current increased. In an unregulated supply, increased load current would mean that the output voltage would decrease because of the drop across any series resistance within the supply. Hence, if the −6 V tends to drop, the swinger of the potentiometer will go slightly more positive than before. Hence the base of Q_1 is slightly more positive. The collector current of Q_1 must decrease, and a smaller drop occurs across R_1, which is the load resistor for Q_1. This causes the base of Q_2 to swing more negative, and turns it on

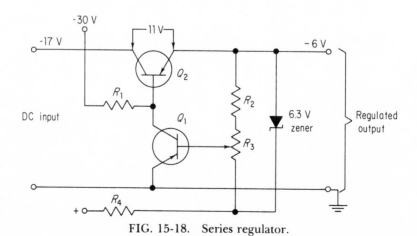

FIG. 15-18.   Series regulator.

harder, which, of course, reduces its internal resistance. Since a larger current is now flowing through a small resistance, the voltage drop tends to remain the same. If the component values are chosen properly, nearly exact regulation can be achieved in this way.

The series regulator has the further advantage that changes in the dc input are also compensated for. Suppose the line voltage increased to the extent of causing the −17 V to increase to −18 V. We should expect the −6-V output to increase to −6.35 V if the supply were not regulated. However, −18 V on the collector of Q_2 means that the emitter of Q_2 will tend to swing slightly more negative. This is reflected as a more negative voltage on the base of Q_1, turning it on harder, and the base of Q_2 is driven more positive. Again, a more positive voltage on the base of Q_2 increases its internal resistance, and the drop across it increases. Now 12 V is dropped across the transistor, and $−18 + 12 = −6$ V is still across the load.

The series regulator is then capable of compensating for changes in input voltage as well as for changes in output voltage. It is, however, more complex.

15-8 POWER-SUPPLY CHARACTERISTICS

It is nearly impossible to obtain absolutely pure, unvarying dc from a rectifier, even though a multiple-section filter is used. There is always some amount of ripple left. This is usually expressed as a percentage of the ratio of the rms value of ripple voltage to the average total voltage output of the filter. This can be written as

$$\% \text{ ripple} = \frac{E_{ac}}{E_{dc}} \times 100$$

For instance, suppose we are analyzing the waveform shown in Fig. 15-19 for ripple content. As shown, the rms value of the ac component is 1 V. The *average* value of the total output is 50 V, so

$$\% \text{ ripple} = E_{ac}/E_{dc} \times 100 = \frac{1}{50} \times 100 = 0.02 \times 100 = 2 \text{ percent}$$

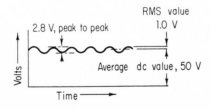

FIG. 15-19. Waveforms used to develop percent ripple.

Another important characteristic of the power supply is the voltage regulation inherent in its design (excluding electronic voltage regulators at this point). The percent regulation is a measure of how the output voltage varies with changes in load current, and can be determined by

$$\% \text{ regulation} = \frac{100 \ (E_{max} - E_{min})}{E_{min}}$$

The maximum voltage output will occur with minimum load current, usually specified as no-load current. The minimum output will occur with full-load current. If, under these conditions, E_{max} is 50 V and E_{min} is 45 V, then for this particular power supply

$$\% \text{ regulation} = \frac{100(50 - 45)}{45} = \frac{500}{45} = 11.1 \text{ percent}$$

With capacitor-input filters the regulation may approach 100 percent, which is very poor. A highly regulated supply as described earlier may have regulation as good as 0.01 percent or better.

Components

The various components of the power supply must have certain ratings to be able to perform their jobs efficiently with a reasonable life expectancy. Although we shall certainly not be designing power supplies, we must be aware of these factors, so that when we are servicing the equipment, we shall be able to substitute intelligently a defective component with as good, or better, replacement.

The transformer, of course, must deliver the proper voltages at a certain value of current. It must be rated for any unusual voltage stresses that exist at any winding, and it must have an adequate volt-ampere rating. (The "V-A rating" of a transformer determines the maximum power that can safely be delivered without overheating.)

The capacitors used in the filter must be capable of withstanding the peak voltage delivered by the transformer. Under no-load conditions, or even light-load conditions, these capacitors can easily charge to the peak voltage. Their dc working voltage must be greater than the highest voltage expected, to avoid breakdown. The value of capacitance is fairly critical too.

The filter choke of course must have the rated inductance in henrys, but equally important is the dc resistance of the winding. If too high, the power-supply regulation suffers, and if too low, the choke is probably larger and more costly than necessary. Also, the choke is usually placed in the high-voltage lead. Its windings must be insulated for the maximum expected voltage, since the iron core is usually at chassis ground.

Finally, the rectifiers themselves must, of course, be capable of passing the necessary current and, additionally, must be rated for the maximum peak inverse voltage (PIV) to be expected. That is, each diode sees a reverse voltage for one-half of each input cycle. The voltage across it must not exceed the breakdown value or the diode will be destroyed. In low-voltage circuits this is less of a problem than in high-voltage supplies, but nevertheless it must be taken into consideration.

Input versus Output

There are occasions when the servicing of a power supply makes use of the relationship of output-to-input voltage and current. This is especially useful if the supply in question uses a choke-input filter of proper design, for then the conditions can be specified more exactly. Unfortunately, specifying the exact output of a capacitor-input filter is quite complex because of the many variables. For instance, if the full-wave-rectified output to the filter is 100 V rms, the output can vary from 90 to 142 dc V. Even with a steady load, the actual output depends upon (1) the line frequency, (2) the size of the filter capacitors, (3) the internal resistance of the supply, and (4) the amount of load current. Thus, not only is regulation poor, but the relationship of output voltage to input voltage is, at best, nebulous.

The choke-input filter, on the other hand, has characteristics that enable us to predict quite accurately the necessary input (ac) to produce a given output (dc). For example, a single-phase full-wave rectifier, shown in Fig. 15-20, will exhibit the voltages and currents as given.

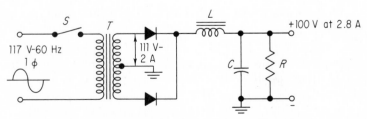

FIG. 15-20.   Example of choke-input-filter circuit.

Assume that the full transformer secondary produces 222 V rms. The dc output will be determined by the following relationship. The actual value of ac available for rectification is one-half of the full secondary voltage, or 111 V.

$$E_{dc} = 0.9 \times \frac{E_s}{2} = 0.9 \times 111 = 100 \text{ V}$$

(Again, 0.9 is the form factor derived from $E_{av}/E_{rms} = 0.636/0.707$.)

If the transformer, the diodes, and the choke drop any appreciable voltage, the output will be less than 100 V by the total of their drops. For simplicity, we shall assume no internal drops, and the dc output is 100 V.

If the transformer is rated at 2 A rms, the dc output current will be

$$I_{dc} = 1.4 \times i_{rms} = 2.8 \text{ dc A}$$

$$(I_{dc} = \text{peak current; so } i_{rms} = 0.707 \, I_{dc}.)$$

Note that these figures apply only to the choke-input filter used with a full-wave rectifier. Any other type of rectifier or filter would require different form factors (0.9 and 1.4).

Precautions

Power supplies are generally quite rugged, and supplies using vacuum tubes especially so. When semiconductors are used, however, certain precautions must be observed, since a crystal diode or transistor can be utterly destroyed in a very few milliseconds. This is particularly true of unregulated supplies using semiconductors.

In the case of a shunt regulator similar to that shown in Fig. 15-17, the load must never be removed while operating. Both the zener and the transistor can be destroyed by even a momentary open in the load.

The series regulator, as shown in Fig. 15-18, must never be short-circuited, or the series transistor will surely be ruined, unless the circuit is designed to prevent this. Modern power supplies are often designed to be short-circuit protected and are then able to withstand a continuous short circuit across the output without damage.

QUESTIONS AND PROBLEMS

15-1   (*a*) In the circuit shown, what is the polarity of E_1 with respect to ground?
(*b*) What is the polarity of E_2 with respect to ground?
(*c*) What is the polarity of E_3 with respect to ground?

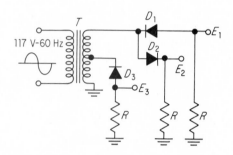

In the following questions, select the correct answer.

15-2   (*a*) If, in the drawing of Question 15-1, the value of E_1 were
100 V, E_2 could be expected to be
(1) greater   (2) same   (3) less
(*b*) If E_1 were 100 V, E_3 could be expected to be
(1) greater   (2) same   (3) less

15-3   In the circuit shown, what is the output voltage as read on a
dc voltmeter? Assume a perfect transformer with a step-down
ratio of 5:1 and no losses in the secondary circuit.
(*a*) 51 V   (*b*) 254 V
(*c*) 18 V   (*d*) 40 V

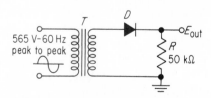

15-4   (*a*) In the circuit shown, the voltage at point *B*, in respect to
A, is
(1) positive   (2) negative

(b) If the transformer has unity turns ratio (primary to one-half secondary), what is the voltage across RL? (Unity turns ratio = 1:1.)

(1) 117 V   (2) 166 V
(3) 111 V   (4) 61 V
(5) 53 V

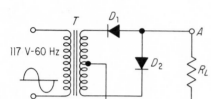

15-5 Which of the waveforms shown is most like that of an unfiltered full-wave rectifier?

(a)

(b)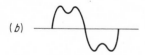

(c)

(d)

(e)

15-6 A disadvantage of the full-wave rectifier is that the output requires the most filtering of any configuration.
(a) True   (b) False

15-7 Is the drawing below correct in every detail?
(a) Yes   (b) No

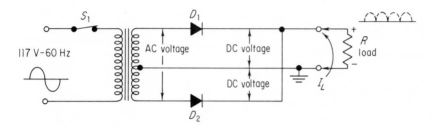

15-8 Is the drawing below correct in every detail?
(a) Yes   (b) No

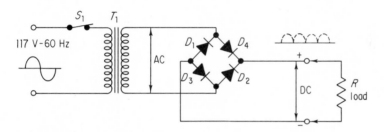

15-9 In the circuit shown, the value of E_1 is
            (1) 10 dc V   (2) 90 dc V
            (3) 90 ac V   (4) 110 ac V
            (5) 55 ac V
     (b) The value of E_2 is
            (1) 99 dc V   (2) 110 dc V
            (3) 55 dc V   (4) 110 ac V
            (5) 90 ac V
     (c) The value of E_3 is
            (1) 110 dc V   (2) 99 ac V
            (3) 90 dc V   (4) 110 ac V
            (5) 55 ac V

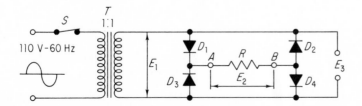

15-10 In the circuit of Question 15-9,
(a) A is more positive than B
(b) B is more positive than A

15-11 (a) In the diagram shown, the polarity as given is
(1) correct   (2) not correct
(b) The maximum possible voltage across R, for the conditions shown, is
(1) 117 V   (2) 234 V
(3) 331 V   (4) 58 V
(5) 165 V
(c) The maximum possible voltage across C_1 for the conditions shown is
(1) 17 V   (2) 234 V
(3) 331 V   (4) 58 V
(5) 165 V

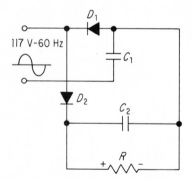

15-12 A certain power supply is rated to produce no more than 1 percent ripple at 35 dc V. When an oscilloscope is used to measure the dc, it is noted that some ripple is seen. Carefully measuring the ac ripple yields a 45-mV peak-to-peak reading, and the average dc reads 35 V. Is the power supply within specifications?
(a) Yes   (b) No

CHAPTER 16 TRANSISTOR TYPES

There are so many kinds of semiconductor devices on the market that it is impractical to discuss them all. To acquaint the reader with a few devices that are different from those that have been discussed up to this point, a brief description of several is given. The reader who wishes to expand upon these brief descriptions should consult the bibliography, which lists several sources of information.

The unijunction transistor, the field-effect transistor, and the silicon-controlled rectifier have been chosen as representative of these other semiconductor devices. Finally, a few words describe the four major ways in which transistors are made, to allow one to better visualize some of the mechanical differences between transistor types.

16-1   THE UNIJUNCTION TRANSISTOR

The unijunction transistor is a three-terminal device that behaves very differently from a regular junction transistor. As the name implies, it has but one junction, with two base leads and one emitter lead. This is shown in Fig. 16-1 both schematically and diagrammatically.

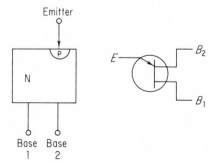

FIG. 16-1.   Unijunction transistor.

A chip of N-type silicon is used as the base material, and the P-type emitter is formed on it, creating a PN junction. One of the base leads (B_2) is closer to the emitter than the other.

The resistance between the two base leads is typically 6 to 8 kΩ. In a normal circuit, the resistance from the emitter to either base is variable, depending upon the circuit and the applied voltages.

Base 1 is usually placed at circuit ground, and the circuit power supply V_{BB} is applied to base 2. The emitter, then, becomes the input connection, while base 2 provides the output. Other connections are also possible. However, the unijunction transistor is not capable of linearly amplifying a signal. Its main uses depend upon its "negative-resistance" characteristic. Some of these uses are:

1.   SCR triggering
2.   Oscillator
3.   Timing
4.   Pulse generator
5.   Bistable circuits
6.   Sensing circuits

In all these instances, use is made of the fact that as a signal is injected, the transistor goes from the off to the on condition, or vice versa. There is no useful in-between area of stable operation.

A characteristic curve for a typical unijunction transistor is shown in Fig. 16-2. The ordinate of the graph is labeled V_E (emitter voltage) and increases in the upward direction. The abscissa is emitter current and increases to the right. Note that in the negative-resistance region, as emitter current increases, emitter voltage *decreases.* This is exactly the opposite of what we might expect. Any device that has smaller voltage drop as the current is increased is said to possess

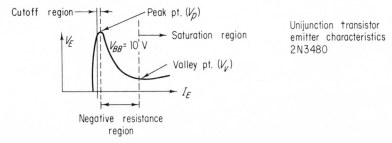

FIG. 16-2.   Unijunction transistor 2N3480. (*By permission of Motorola Semiconductor Products, Inc.*)

negative resistance. Any device exhibiting negative resistance is capable of regeneration, and thus is capable of oscillating.

The region to the left of the peak point is called *cutoff*, while the region to the right is called *saturation*.

Some of the more important terms used in conjunction with unijunction transistors are listed and explained below:

1. Interbase resistance R_{BB}. The interbase resistance is the ohmic resistance measured between base 1 and base 2 with the emitter open.
2. Intrinsic stand-off ratio η. The intrinsic standoff ratio is a number less than 1 that represents the amount of applied voltage necessary to fire (turn on) the transistor. Mathematically,

$$\eta = \frac{V_P - V_D}{V_{BB}}$$

Typical values of η are from 0.4 to 0.8.
3. Peak-point current I_P. The peak-point current is the emitter current at the peak point. This is the minimum current needed to turn on the transistor.
4. Peak-point emitter voltage V_P. This is simply the emitter voltage at the peak point.
5. Emitter reverse current I_{eo}. This is equivalent to I_{cbo} in a conventional transistor. It is measured between base 2 and the emitter, with base 1 open.
6. Valley voltage V_V. This is the emitter voltage at the valley point.
7. Valley current I_V. This is the emitter current at the valley point.
8. Diode voltage V_D. This is the drop across the PN junction, equal to

$$V_D = V_P - \eta V_{BB}$$

Unijunction-Transistor-Circuit Example

A typical oscillator circuit will serve to show how the device can be effectively used. Such a circuit is shown in Fig. 16-3. This is a basic relaxation oscillator.

At the beginning of a cycle of operation, say, point A on the waveforms, the capacitor begins to charge toward $+V_{BB}$ through R_3. At this time the transistor is off and does not influence circuit action. This is shown at B. When the emitter voltage reaches the peak-point value V_p, the unijunction transistor turns on, and the capacitor discharges into R_1 and the emitter, shown at point C.

When the capacitor voltage drops very low, the transistor turns off, and the cycle begins over again.

This circuit could be used for timing purposes, as a pulse generator, as a trigger circuit, or as a sawtooth-wave generator.

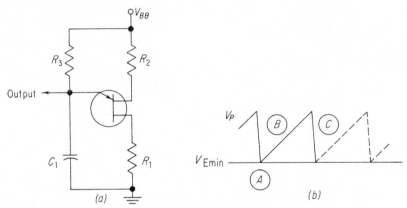

FIG. 16-3.   (a) Unijunction transistor circuit; (b) output waveform.

16-2   FIELD-EFFECT TRANSISTORS

The field-effect transistor (FET), as used today, is a relatively recent development. Only in the last few years, since about 1960, has it occupied a place of prominence in the industry. Its usage is increasing of late, and it is reasonable to assume that one might encounter such a device in various circuits. Therefore the well-informed technician should be familiar with it.

There are two major types of field-effect transistors: the junction field-effect transistor and the insulated, or isolated, gate field-effect transistor, abbreviated JFET and IGFET, respectively. To begin to understand how these devices work, consider Fig. 16-4. Figure 16-4a represents a wafer of silicon that acts much like an ordinary resistor. The two connections are labeled the "source" and "drain." The device as shown would conduct current if a voltage were impressed upon the terminals, the amount of which would be proportional to the value of voltage. In Fig. 16-4b two additional areas have been added: gate 1 and gate 2. These areas have been doped with P material, while the main bar of silicon is N type. Thus each causes the formation of a junction. As with all PN junctions, a depletion region is formed at the junction, as in Fig. 16-4c. The region between the gates is called the *channel*, which is a part of the main bar, or wafer, of silicon, known as the substrate. Any current flowing through the device must pass through the channel.

If current is caused to flow, as shown in Fig. 16-4d, the shape of the depletion region is altered. As current is increased to a large

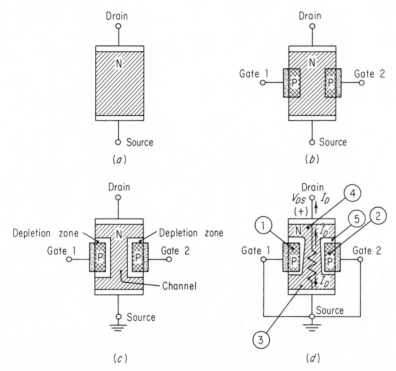

FIG. 16-4. Development of junction-field-effect transistors. (*From Application Note AN 211, Motorola Semiconductor Products, Inc.*)

value, the depletion regions meet and restrict any further increase in current. The applied voltage that just causes this effect is called the *pinch-off voltage* V_P.

Since it is difficult to produce the gates on both sides of a bar as shown in Fig. 16-4, a single-ended method is used to produce the practical field-effect transistor. Figure 16-5 shows such a structure, and the various parts are numbered to correspond with the numbers on the drawing of Fig. 16-4d. Area 2 is the substrate, which forms gate 2, and is the main part of the structure. Area 5 is the N-type channel that is diffused into the surface of the block of silicon. Finally, area 1 is formed by diffusing a P-type material to form gate 1, as shown. Electron flow, from source to drain, is also shown, and note that it must pass through the channel. A depletion region surrounds the channel at every PN junction. This is the mechanism by which the drain current can be made to vary according to a signal. If gate 1 is caused to vary its voltage, the depletion regions will vary, and the current flow will also vary in step with the input at gate 1.

The electrical symbol for the JFET is shown in Fig. 16-6. As

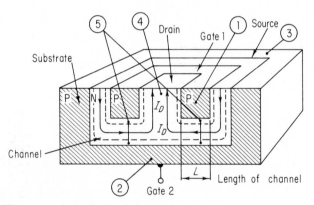

FIG. 16-5. Single-ended configuration. (*From Application Note AN 211, Motorola Semiconductor Products, Inc.*)

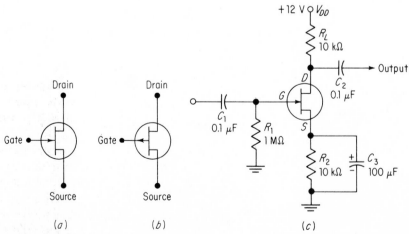

FIG. 16-6. Junction FETs: (*a*) N-channel device; (*b*) P-channel device; (*c*) JFET Amplifier.

shown, the gate connection is used as an input, with the drain as the output. The source and the substrate connections are often connected together internally.

The insulated gate field-effect transistor is constructed somewhat differently from the junction counterpart. Figure 16-7 shows how the IGFET is developed. In this case, two separate N-type regions are diffused into the substrate to form the insulated source and drain. Then the entire surface is covered with silicon dioxide, which is an excellent insulator. Openings are made in the oxide layer to allow contact to be made between the source and drain connections and the N channels. Next a metal covering is laid over the center section, which becomes gate 1. There is no physical connection from gate 1 to the semiconductor proper. Thus the metal connector, the oxide insulator, and the channel just beneath form a capacitor. This is a most important idea, because any voltage impressed upon gate 1 can affect the transistor *only* through this capacitance.

Because the source and the drain are isolated by the substrate, the drain current I_D is essentially zero with zero gate voltage. This is true because the internal junctions between the source and the drain act like back-to-back diodes. With any applied voltage, one of them will be back-biased. In order to cause current to flow through the

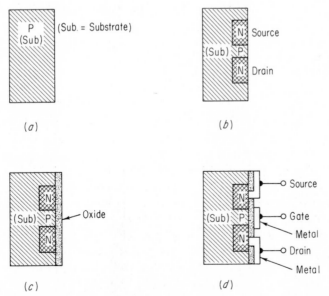

FIG. 16-7. Development of insulated-gate FET. (*From Application Note AN 211, Motorola Semiconductor Products, Inc.*)

device, it must be "turned on." This is accomplished by applying a positive voltage to the gate with a normal positive voltage applied to the drain. As gate 1 is made positive, an "induced" channel is formed, as shown in Fig. 16-8. As the upper plate of the effective capacitor becomes more positive, the lower plate, just below the oxide layer,

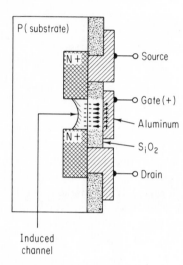

FIG. 16-8. Channel enhancement. (*From Application Note AN 211, Motorola Semiconductor Products, Inc.*)

becomes more negative. The accumulated electrons become carriers, and will now allow current to flow from the source to the drain.

The number of electrons available to carry current is a function of the gate voltage. The drain current can therefore be made to vary in proportion to the applied signal voltage at the gate. Increasing the gate voltage is said to "enhance" the drain current.

One advantage of this kind of transistor is that its input impedance is very high, since the gate acts much like a capacitor. The electrical symbol for the IGFET is shown in Fig. 16-9.

Comparing the two types of FETs just described, we find that the first of these, the JFET, operates fully on with zero gate 1 voltage, and can only be turned "more off." This is called the *depletion* mode of operation. On the other hand, the IGFET is normally off, and can only be turned "more on" by the application of gate voltage. This is called the *enhancement* mode of operation.

The depletion mode refers to the decrease of carriers in the channel due to an increase in positive gate voltage. "Enhancement" refers to the increase of carriers in the channel due to the increase of positive voltage on the gate.

A third type of field-effect transistor is also possible. This is known as the *depletion-enhancement* type. In this case, the zero-gate-voltage drain current is intermediate between full on and full off. Hence both depletion and enhancement of the carriers are normal for this device. This result is accomplished by diffusing a thin N

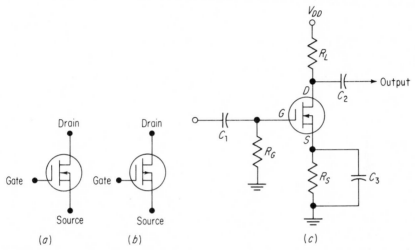

FIG. 16-9. Insulated gate FETs: (*a*) N-channel IGFET; (*b*) P-channel IGFET; (*c*) N-channel IGFET amplifier.

channel between the source and the drain of a regular IGFET just below the oxide layer. This yields a layer that is conductive, and the carriers in the conductive layer can be augmented or decreased by the application of gate voltage.

From a circuit standpoint (Figs. 16-6*c* and 16-9*c*) the field-effect transistor can be seen to be biased much differently than the bipolar transistor discussed thus far. The junction field-effect transistor is a normally-on device. That is, with zero gate-to-source voltage, the unit passes some maximum amount of current. As the gate-to-source junction is caused to become more reverse-biased, current decreases. With some large amount of reverse voltage, current drops to zero. Hence, the JFET is biased in a manner not unlike a vacuum tube. Since current flows as long as voltage is applied to the drain-source circuit, the resistor in the source lead causes a voltage drop, to reverse-bias the gate-source junction. For example, in Fig. 16-6*c*, R_2 performs this function, with its upper end more positive than ground. The gate is returned to ground, so the channel is caused to be more positive than the gate, thus reverse-biasing this junction. Note the large return resistor in the gate lead. A 1-MΩ resistor is typical for FET circuits, but would be much too large for use in a bipolar transistor circuit. To avoid severe degeneration and to increase the overall voltage gain the source resistor is bypassed. The foregoing description is equally valid for the circuit shown in Fig. 16-9c, if account is taken of the increased input resistance of the IGFET. The gate-return resistor might be on the order of 10 MΩ in this example, if such a large value were useful.

A word of caution: when handling IGFETs, allow for the fact that they are easily destroyed by static discharge from the body. Keeping all transistor leads shorted together until placed in a circuit will prevent this kind of damage.

16-3 THE SILICON-CONTROLLED RECTIFIER (SCR), OR THYRITE

The silicon-controlled rectifier is a special case where the properties of semiconductors are used to produce a special effect. The SCR is a solid-state device having certain properties quite similar to a thyratron gas tube. That is, the control element, called the *gate*, can maintain sufficient control to keep the device off (open-circuited) for any period of time, provided the device is already off. By injecting a current into the gate, the device is turned on, and will conduct current. Once the conducting state is reached, the gate loses its ability to control current flow, and can turn the main device neither off nor further on. If now the current path is interrupted, the gate can once again regain its ability to keep the device off, even though an otherwise complete circuit is reestablished.

An SCR is made of four alternate layers of P- and N-type silicon. Its physical construction is symbolically illustrated in Fig. 16-10. The SCR is seen to be the equivalent of two separate transistors: an

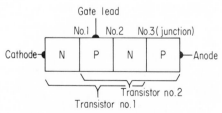

FIG. 16-10.   SCR (PNPN switch).

NPN unit and a PNP. The way it operates can best be appreciated if we split the two center sections in two pieces, as shown in Fig. 16-11b, and connect them with ordinary wire.

With the voltages applied as shown in Fig. 16-11b, and with the switch in the gate lead as shown, little or no current flows in the anode

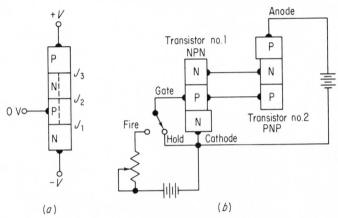

FIG. 16-11.   Equivalent circuit for an SCR.

circuit. Inspection of the three junctions reveals the fact that junction 2 (J_2) is reverse-biased; so no appreciable current can flow, since junction 1 is not forward-biased.

The base of the NPN transistor is not forward-biased because of the switch position, and the NPN unit is nonconducting. Any current flow in the anode circuit (PNP unit) must flow through the NPN part, and so the anode current is essentially zero. (A very small tempera-ture-dependent current similar to the I_{cbo} in a transistor will exist.)

If now the switch is transferred with a large resistance in series, a small base current will flow in the NPN unit. (The circuit is redrawn in Fig. 16-12a.) The collector current will then be beta times the NPN base current. This current is injected into the base of the PNP unit, and will appear at the emitter of the PNP unit with an amplitude of $(\beta' + 1)\,(\beta)$ times the original base current (β = NPN, β' = PNP). If we gradually reduce the resistance in the NPN base, more and more current will flow in the base circuit. When the initial base current is made large enough so that the current in the emitter of the PNP part is exactly the same value as the original base current, the loop current gain is 1. Now if the base of the NPN part is disconnected from the battery supplying base current, the device will *continue to provide its own base current.* Both the NPN and the PNP parts will go into saturation. A large current will flow in the anode circuit, the value of which is dependent upon the load resistance and the applied voltage.

Since the device is now supplying its own internal base currents, what we do to the external NPN base circuit (gate) can in no way

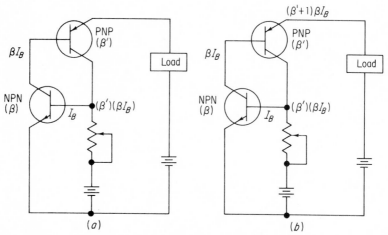

FIG. 16-12.   SCR current relationships.

affect the conductivity, and the gate is said to have lost control. Anode current can be stopped only by opening the anode circuit or reducing the anode current below the point of regeneration. Once anode current stops, we can again regain control by stopping gate current. If anode voltage is again applied, no anode current will flow until the gate current is again made large enough to produce regeneration.

Some of the terms used in describing SCRs are listed below for handy reference.

I_f	Forward anode current. The value of anode current through the device when on.
$I_{f(off)}$	Forward off current. The value of anode current through the device when in the off condition.
I_g	Gate current.
I_h	Holding current. The minimum anode current required to sustain the on condition.
I_r	Reverse current. Any current through the device when negative voltage is applied to the anode.

V_{bf} Forward breakover voltage. Anode voltage that will cause the rectifier to switch to the on state, with no gate current.

V_f Forward voltage. The voltage drop between the anode and cathode at any specified forward current, when the device is on.

$V_{gr\ (rated)}$ Gate voltage, reverse. Maximum allowable reverse voltage applied to the gate junction.

V_r Reverse anode voltage. Any negative value of voltage applied to the anode.

$V_{r\ (rated)}$ Maximum inverse voltage allowed. To exceed $V_{r\ (rated)}$ would cause entry into the avalanche region, and the device would go on, even though no I_g were present.

The schematic diagram of an SCR is shown in Fig. 16-13, with an appropriate circuit that might be used to, say, energize the relay *RE*.

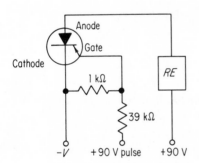

FIG. 16-13. SCR schematic circuit.

The purpose of the voltage divider is to reduce the 90-V pulse to a value suitable for application to the gate. With a 90-V pulse at the input [1 ms (millisecond) duration or longer], slightly more than 2 V will be applied to the gate and will be sufficient to turn on the SCR.

16-4 TRANSISTOR CONSTRUCTION

Transistors are made in a large variety of types, each slightly different from the other, but all having the same general characteristics. Most transistors can be classified as one of four basic types: (1) alloy transistors, (2) grown-junction transistors, (3) mesa transistors, and (4) planar transistors. Each of these can be further subdivided into one or more related types. It will not be necessary to delve into the many subtleties of transistor construction, but we shall find a need for a brief discussion of the major types, with their overall characteristics.

The Alloy Transistor

The alloy transistor is made by alloying metal into opposite sides of a thin piece of semiconductor, one side becoming the emitter, and the

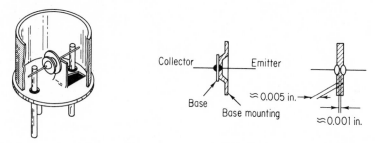

FIG. 16-14.   Alloy transistor.

other side the collector. In order to produce alloy transistors with reasonable uniformity, very close control of such factors as base-pellet thickness, quantity of metal to be alloyed, area of contact, and alloying temperature must be achieved. Because of the difficulty of maintaining consistent base widths, an upper limit on the highest operating frequency is established.

The alloy transistor, then, does not have especially good high-frequency characteristics. It does, however, exhibit excellent saturation resistance, on the order of 1 to 5 Ω when operated in saturation. Figure 16-14 shows the typical construction of an alloy transistor.

Variations of the basic alloy technique are the microalloy transistor, which has better high-frequency characteristics, and the micro-alloy diffused transistor, which is still better in the high-frequency region of operation.

Grown-Junction Transistors

This unit differs from the alloy transistor in that the junctions are created during the crystal growth, instead of alloying each junction after the crystal is grown. The growing crystal is slowly withdrawn from the melted semiconductor, and the melt is alternately doped N and P. Thus 30 or more pairs of junctions are created in a single crystal, each pair of junctions being sawed from the crystal and diced into several hundred individual pellets. Each pellet is a complete transistor, requiring only connecting leads and appropriate mounting (see Fig. 16-15).

The advantages of a rate-grown transistor are the desirable cost factor and mechanical ruggedness. The disadvantages are a limited frequency range and the difficulty of attaching a wire to the base region, which, ideally, is on the order of 0.001 in. (inch).

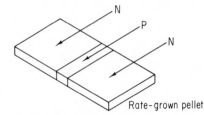

Rate-grown pellet

FIG. 16-15. Grown-junction transistor.

A variation of this technique is the grown-diffused transistor which exhibits a higher-frequency operation.

Mesa Transistors

The mesa construction overcomes many of the disadvantages of grown-junction and alloy transistors. The thinness of the base region does not affect the mechanical reliability as in the case of the other types. The use of base diffusion results in excellent control of the base width, making it possible to build transistors with reproducible characteristics. Mesa construction is shown in Fig. 16-16.

The metal stripes are evaporated onto the surface of the diffused pellet. One of these is gold, which produces a nonrectifying contact

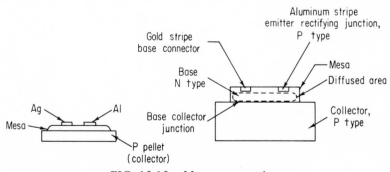

FIG. 16-16. Mesa construction.

and becomes the connection for the base lead. The other stripe is aluminum, which will form a rectifying contact and become the emitter. The diffused layer beyond the limits of the stripes is etched away to reduce the junction capacitance. High-speed switching transistors are often made by this process.

An improvement on the mesa transistor is the epitaxial mesa type of construction. In this case, a thin film is grown on the collector pellet, and the atoms of the film are aligned in a continuation of the original crystal structure. The base region is then diffused into the thin film, and the rest of the fabrication is the same as in the conventional mesa transistor. A considerable improvement in collector-to-base breakdown voltage is achieved, as well as reduced junction capacitance and lower saturation voltage.

Planar Transistor

A silicon planar transistor is made by diffusing the emitter, as well as the base, with no mesa-forming etch used. Use is made of the fact that silicon dioxide can be formed on the surface of silicon pellets to act as a mask to prevent the diffusion of impurities into the silicon. Portions of the SiO_2 are removed to allow the impurity to diffuse into the collector pellet. Then aluminum is deposited on both the base and emitter regions to make electrical contact (see Fig. 16-17). The silicon dioxide layer covers the junctions, preventing the entrance of gases. This is known as *passivation,* since the junctions are in effect sealed away from moisture, etc.

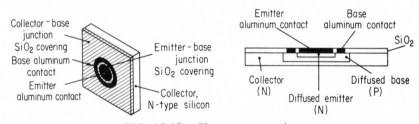

FIG. 16-17.   Planar construction.

QUESTIONS AND PROBLEMS

16-1   Select the correct answer. The symbol shown represents
   (*a*) a junction transistor
   (*b*) a back-to-back diode

(c) a unijunction transistor
(d) a trijunction transistor
(e) an FET

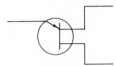

16-2 Select the correct answer. The symbol shown represents
  (a) an FET
  (b) a unijunction transistor
  (c) a trijunction transistor
  (d) a duojunction transistor
  (e) an IGFET

16-3 True or false: The unijunction transistor makes an excellent linear amplifier.

16-4 True or false: Referring to Fig. 16-3, slightly changing the value of R_2 will cause the frequency of oscillation to change.

16-5 Refer to Fig. 16-4. Discuss the reason for the generation of the depletion region.

16-6 Refer to Fig. 16-4. Discuss the development of the effect of the pinch-off voltage on the source-drain current.

16-7 True or false: The SCR (PNPN switch) is controlled in the same general way as a conventional transistor. (That is, it is turned on and off by the proper application of the proper signal to the gate lead.)

CHAPTER 17 INTEGRATED CIRCUITS

Because integrated circuits are used so widely, it is felt that the entire subject is worthy of separate and more or less complete coverage. However, because of limited space, this discussion is, in certain areas, somewhat superficial. Nevertheless, it is hoped that the following information will allow the serious student to fully appreciate integrated circuits and to be able to work on and around them with greater confidence.

17-1 INTRODUCTION

Integrated circuits (ICs) are being used in an increasing number of applications, and it is expected that as time goes by, their use will become much more widespread. They are being used in nearly every type of commercial and military equipment where reduced weight, smaller size, and better reliability are required.

Microelectronics is the art of compressing more and more components in a smaller and smaller volume (package). In earlier days, using vacuum tubes and associated components, about 5000 components could be packed in a square foot of space. Later, when transistors became firmly established, on the order of 100,000 components could be mounted in the same space. Now, however, using integrated circuits, nearly 10 million components might be found in a 1-ft^3 volume. It becomes apparent, then, that such a saving in space, weight, and very often cost can be quite considerable.

Some of the most obviously useful applications for ICs are in space vehicles (or any extraterrestrial vehicle), computers, portable hand-carried equipment, etc. Many other applications come to mind with very little thought.

Integrated circuits have provided the means to reach a solution to one of the greatest problems ever encountered by the electronic design engineer. This problem is described as the *tyranny of numbers,* which refers to the ever-increasing complexity of modern-day equipment and the consequent multiplication of individual parts. As machines become ever more complex, the chance of machine failure due to component malfunction increases by leaps and bounds.

One of the advantages of integrated circuits lies in the lack of individual connections that in the past would have been installed by hand. Many of these connections are an integral part of the chip and as such are removed from the foibles of human frailty. By their very nature, integrated circuits are inherently reliable, consisting of a number of components, both passive and active, inseparably bonded together into an integral unit. This greatly increases total system reliability. Adding to this inherent reliability is the passivation process, which seals each part of the integrated circuit not only from all other parts, but from the outside world as well.

Another significant advantage is the extremely low cost per unit relative to the number of functions on a given unit. As will be seen, up to 1500 identical circuits are processed simultaneously on a single wafer, which is on the order of 1 to $1\frac{1}{2}$ in. in diameter. Because several hundred wafers can be processed at one time, the cost per circuit is very small. However, this does not necessarily mean that all integrated circuits are of inconsequential cost. After the chip is separated from its neighbors on the wafer, it must then be thoroughly tested and mounted in a suitable enclosure, and leads must be attached. These steps in the manufacturing process cannot be accomplished en masse, and from this point on, much hand labor is involved. It is here that most of the ultimate cost occurs.

Because of minor imperfections in the crystal structure, the possible infusion of dust and other contaminants, as well as other considerations, not every circuit measures up to standards. Some percentage of every batch must be discarded at some point in the overall process. This, of course, also has a large bearing on the ultimate cost. Nevertheless, disadvantages notwithstanding, the integrated circuit is here to stay, at least until an even more revolutionary process comes along to supplant it.

It must not be assumed that the integrated circuit has no disadvantages at all. For example, one limiting factor in the application of ICs is the relatively low power dissipation of the device. For many applications, this is not a severe drawback, but in others it is. Also, the integrated circuit finds greatest use in applications where there is considerable redundancy, that is, where the same circuit can be used over and over again, as in the case of a digital computer. In equipment that uses highly distinctive circuitry, ICs may not provide the most economical approach.

There are four basic techniques used in the manufacturing of ICs. These are the *hybrid, thin-film, monolithic,* and *compatible* circuits. The hybrid circuit is one in which separate component parts or dice (transistors, resistors, etc.) are attached to a ceramic base and connected to each other by means of either wire bonds or a metallization pattern. This construction is very similar to conventional means except that it is enclosed in a container perhaps $\frac{3}{8}$ in. in diameter and $\frac{1}{4}$ in. in height. A thin-film circuit consists of microscopically thin films of material deposited on a ceramic base to form the passive components. Then active components must be added in discrete form to these thin-film networks. With the monolithic (single-stone) technique, all parts of the circuit, including transistors, diodes, resistors, and capacitors, are formed within the single wafer of silicon. Finally, the compatible circuits are those which have the active components formed within the chip of silicon but have the passive components deposited by thin-film techniques on top of the insulating layer that covers the active components.

In the following discussion, we shall be concerned primarily with the monolithic integrated circuit since this is by far the most widely used method.

17-2 MANUFACTURING PROCESSES

While it is not necessary to know how ICs are made in order to apply them, certain advantages accrue when at least a cursory investigation is made. At the very least, some of the processes used are enlightening and quite interesting, being unique in the realm of electronics.

To introduce this subject, refer to Fig. 17-1. The object at the top of the figure is a bar (ingot) of pure silicon that has been produced, or grown, by "pulling" a seed crystal from a mass of molten silicon under very controlled conditions. The ingot of pure silicon represents the first stage in IC production. It is about 6 to 8 in. in length and between 1 and $1\frac{1}{2}$ in. in diameter. The next step in the manufacturing process is to cut the ingot into very thin slices, or wafers, that are on the order of 12 mils (0.012 in.) thick. Many, many wafers can be sawed from each ingot. The wafers are extremely brittle and will break quite readily, and hence they must be handled carefully. Each wafer is then lapped to about 6 mils, using a very fine grit abrasive, and chemically etched to form an extremely flat and smooth surface.

At this point in the overall process, the wafers are subjected to the epitaxial growth of a layer of silicon over the face of the wafer that will eventually hold the circuits. The need for the epitaxial growth

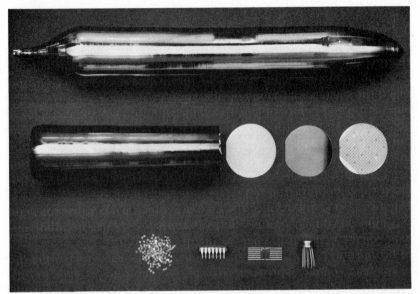

FIG. 17-1. Various steps in the manufacture of ICs. (*Courtesy Fairchild Semiconductor.*)

lies in the fact that in the following processes, which are accomplished by diffusion, no more than three layers can be produced satisfactorily. The added layer produces an exact extension of the underlying crystal structure. That is, the exact structure of the atoms in the original crystal (the wafer) is duplicated in the epitaxial layer, which becomes simply an extension of the substrate about 10 μ in depth ($10\ \mu = 10 \times 10^{-6}$ m).

In practice, the epitaxial layer is a doped structure and can provide either a P- or an N-type layer. It is grown in a furnace, with a carefully controlled temperature, containing an atmosphere of silicon, chlorine, hydrogen, and a phosphorus compound for N-type doping or a boron compound for P-type doping. A typical structure is shown in Fig. 17-2. The epitaxial layer is grown upon the substrate (body of the wafer), and the transistor junctions are formed in this layer by the process of diffusion, to be described subsequently. While transistors can be formed on the substrate with no epitaxial layer, the overall characteristics of the device (BV_{cbo} and $V_{CE,\text{sat}}$, in particular) are greatly improved. The formation of the junctions in the epitaxial layer affords a simple and effective way to produce a compromise between these conflicting requirements.

Next, the wafers are placed in a furnace containing an oxygen atmosphere at 1200°C. The oxygen penetrates the surface of the silicon and combines chemically with the atoms of the crystal lattice, forming silicon dioxide, a stable, inert glass. The dioxide layer envelops the wafer and *passivates* the wafer surface, which greatly reduces the possibility of contamination. The wafer is now ready to

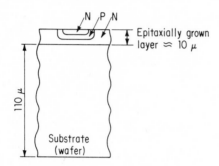

FIG. 17-2.   The epitaxial layer.

begin the processes that will ultimately lead to the final integrated circuits, up to 1500 of which may be formed on the single wafer.

Before the next processes can be accomplished, several other steps must be taken in preparation for the actual formation of the ICs. First, the electrical design must be finalized so that when the ICs are finished, it is known for certain that the circuit will have the required electrical characteristics. Then, the basic breadboard is trans-

FIG. 17-3.   Typical IC artwork.

formed into the required artwork drawings, one of which exists for each step of the process. Each drawing represents the areas on the chip that are to be operated upon during that particular step. The artwork is drawn first about 30 × 30 in. to ensure the highest possible accuracy. Figure 17-3 illustrates a typical piece of artwork for a very

simple device. In practice, these can become much more complex.
Each panel is then reduced by photographic means about 500 times.
Then, a small glass plate, identical in size to the wafer, is treated with
photosensitive material, and the image is exposed repeatedly across
the face of the plate. Thus, as many as 1500 identical images may be

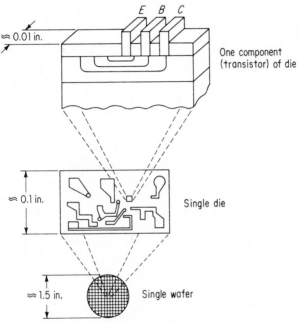

FIG. 17-4.   Location of a single transistor relative to completed wafer.

produced on a single plate, which is used as a master to expose the
silicon chip. With all master glass plates ready, the silicon wafers are
ready to be transformed into completed integrated circuits.

Because of the complexity of a typical IC, it is not feasible to show
the complete process. Figure 17-4, however, illustrates the portion of
a typical circuit that we shall attempt to depict. Taking any of the complete circuits on the wafer and separating it from its neighbors, one
single component (a transistor) from the circuit will be dealt with to
illustrate in a very simplified manner how the IC is built up, step by
step. Throughout the following discussion, keep in mind that the
individual steps that are correlated with the various parts of the transistor are, *at the same time,* producing *all* collectors, *all* bases, and *all*
emitters, as well as other components, such as resistors.

The passivated wafer begins to take the form of an integrated circuit when the collector cutout is made. The wafer is coated with a layer of photosensitive material and then is exposed to light through the glass mask that has been prepared to process all collector regions at once. The portions of the wafer that are exposed to light become hardened and in the subsequent rinse remain on the disk, with all other areas having a coating that is easily removed. Next, a hydrofluoric acid etch is used to remove the silicon dioxide from those areas not protected by the layer of photosensitive material (photoresist). In this manner, windows in the dioxide layer are produced to allow the next step, collector diffusion, to take place.

The wafer is then placed (along with many others) in a furnace whose atmosphere contains an N-type dopant. As the temperature is raised to about 1200°C, the dopant begins to diffuse into the surface of the wafer. That is, the atoms of the impurity are so violently agitated that they bombard the surface. Many of them penetrate into the inner regions and become a part of the silicon structure. This produces a highly doped N-type region that is the collector. The control of the depth of penetration is easily accomplished by accurate temperature control and timing of the diffusion process, and hence the collector can be constructed with exactly the desired characteristics. Figure 17-5 shows a portion of the wafer prior to the diffusion process, while

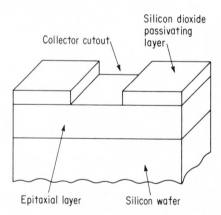

FIG. 17-5. Collector cutout prior to diffusion.

Fig. 17-6 indicates the same section after the collector has been formed.

The next step is to form another layer of silicon dioxide over the entire layer (not illustrated) and a second window exposed and etched over the previous one. This window is smaller than before and is illustrated in Fig. 17-7. The wafer is again placed in the furnace, and this time a P-type dopant is used to diffuse the base region. This, of course, forms a PN junction that is to become the collector-base junction. The diffusion is carefully controlled to allow exactly the correct degree of penetration.

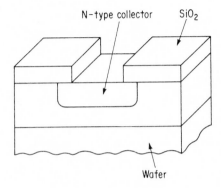

FIG. 17-6.   Collector region after diffusion.

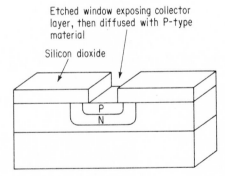

FIG. 17-7.   Forming the base circuit.

Once again, a layer of silicon dioxide is formed over the entire structure in preparation for the emitter diffusion. A still narrower window is etched, after which the wafer is again subjected to a furnace containing the N-type dopant. The formation of the emitter is shown in Fig. 17-8, where it is seen that the three areas are clearly defined and each extends up to the surface of the chip. A new silicon dioxide layer is formed over the wafer to prepare it for the application of the leads, which must be formed so as to connect all necessary areas to each other. Now, many new windows are made, carefully aligned with the proper area, to allow the formation of the metallized leads.

Typical windows for this stage are shown in Fig. 17-9. One method of doing this is the Metal-Over-Oxide process,[1] which is accomplished by evaporating metal onto the surface of the silicon wafer. Aluminum is literally boiled from a hot tungsten filament, depositing the metal in a thin, even coat over the entire wafer surface, as illustrated in Fig.

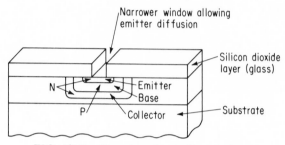

FIG. 17-8.   Forming the emitter region.

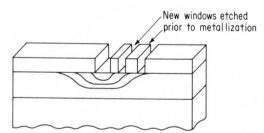

FIG. 17-9.   Preparation of cutouts for metallization.

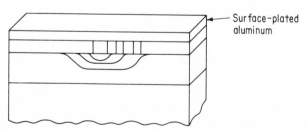

FIG. 17-10.   Forming the interconnection paths.

17-10. One final photoetching process is done at this stage to remove the aluminum over areas where it is not desired. The net result is illustrated in Fig. 17-11, where it is clearly shown that the leads are

[1] Metal-Over-Oxide is a patented Fairchild process.

firmly connected to the three parts of the transistor. The wafer on the right side of Fig. 17-1 shows a completed unit at this stage, with an enlargement of this wafer shown in Fig. 17-13.

The wafers are now ready for final processing. Each wafer consists of up to 1500 individual ICs, each of which may contain several dozen (or possibly several hundred) individual components. By means of rather complex equipment, each circuit on an individual wafer is now tested automatically. Each wafer is inserted in a step-tester having microscopically fine probe tips prepositioned to contact the pads on the periphery of the individual circuits. The wafer is stepped to a position to allow the probe to be lowered, contacting the IC. The probes are connected to an automated tester that evaluates the proper electrical function of the circuit. If the circuit malfunctions, it is marked with dye and will later be destroyed. If it is satis-

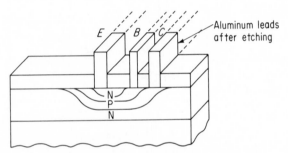

FIG. 17-11.   The completed transistor.

factory, the probes are lifted, the entire wafer is automatically moved so that the next circuit is in position, and the above process is repeated. Thus, every circuit is rapidly tested (Fig. 17-12) under identical conditions.

Next, a diamond-tipped tool scribes a fine line between each circuit, which will allow each to be separated from all others. Each circuit, now called a *die*, is a complete and functioning device. A complete wafer, containing several hundred circuits, is shown in Fig. 17-13 prior to dicing. Figure 17-14 illustrates the size of several typical chips relative to a common paperclip.

However, these dice cannot be used as shown, of course. In order to be able to be connected to the outside world, the unit must be mounted in a suitable container that will allow the connecting leads to be installed. Typical mounting arrangements are shown in Fig. 17-15. Also, Fig. 17-1 shows several possible arrangements. The pads on the chip that were formed during the final metallizing process are bonded to very fine gold wires that connect to the outer leads, by which the device will ultimately be connected in its permanent place. Completed ICs are shown in Fig. 17-16a and b in a closeup view. This particular unit (Fig. 17-16b) has 24 transistors, 6 diodes, and 36 resistors incorporated in its circuit, all on a chip about $\frac{1}{20}$ in. square.

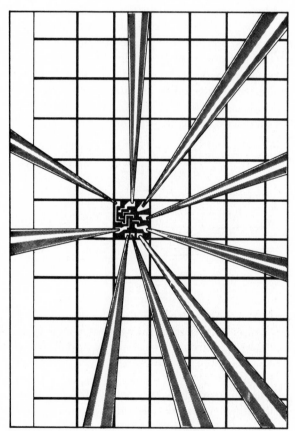

FIG. 17-12.   Automated testing of the chip.

One serious problem encountered in monolithic ICs is that of isolating the various components from one another. The reader may have noticed that if two transistors are constructed side by side, the two collectors are, in effect, connected through the rather low resistance of the epitaxial layer. This, of course, must be avoided at all costs. By doping the epitaxial layer with P-type material (for NPN devices), this effect can be eliminated almost entirely. Figure

FIG. 17-13. Completed wafer containing many individual circuits before dicing. (*Courtesy Fairchild Semiconductor.*)

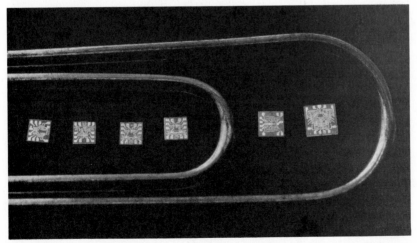

FIG. 17-14. ICs relative to an ordinary paperclip. (*Courtesy Fairchild Semiconductor.*)

FIG. 17-15.   Integrated circuit in a TO-5 case, showing lead posts and lead bonds. (*Courtesy Fairchild Semiconductor.*)

FIG. 17-16.   Completed ICs mounted in (*a*) plug-in receptacle and (*b*) 14-pin (DIP) container.

17-17a shows the effect with no provision for isolating the transistors on the same chip. The resistance from collector to collector may be very low in value and will certainly prevent the collectors from acting independently. In Fig. 17-17b the internal structure of the monolithic chip is shown, which indicates the manner of isolation. In this instance, the collectors and the epitaxial layer now form two diode junctions, which, if the proper connections are made, can be caused to be reverse-biased at all times, thus effectively isolating the two collectors. The epitaxial layer is connected to the most negative point in the circuit (this is often ground), which will cause the diodes (Fig. 17-17c) to always be reverse-biased since all voltages in the circuit will be more positive than this.

One further problem is illustrated in Fig. 17-17d. The astute reader will have noticed that a second transistor exists at the location

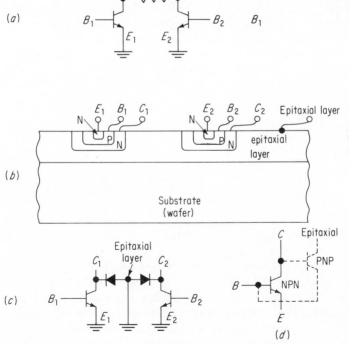

FIG. 17-17. One method of isolating active component parts in a monolithic IC.

of each NPN unit. The base of the NPN unit, along with the collector and epitaxial layer, form a PNP transistor, and this so-called "parasitic" transistor can interfere with normal operation. As before, the connecting of the epitaxial layer to the most negative potential will never allow this parasitic transistor to become forward-biased since its base will always be more positive than the emitter. Certain manufacturing processes (notably gold doping) can reduce the beta of the PNP parasitic transistor to a value such that transistor action is negligible. The parasitic junctions, then, act simply as diodes, which can easily be made to be always reverse-biased.

The above method of component isolation is probably the most widely used, but other ways of accomplishing the same thing are possible. One such method is to layer the substrate with a thick coating of silicon dioxide, and then to etch tiny pockets in the insulating layer. Each active component is then constructed by the usual means completely within the confines of the pocket, thus effectively isolating each transistor, each diode, etc.

The formation of passive components (resistors and capacitors, primarily) is accomplished by diffusion techniques very similar to those used in making transistors. Diodes, on the other hand, are formed by simply connecting the base to the collector of a transistor, as illustrated in Fig. 17-18. The emitter-base junction is used rather

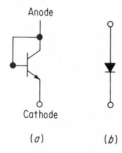

Anode

Cathode

(a)

(b)

FIG. 17-18. (a) Diode formed by connecting the base to the collector; (b) usual schematic representation.

than the collector-base junction because the inherent capacity of this junction is less. Hence the circuit is usable at higher frequencies.

A typical silicon dioxide capacitor is illustrated in Fig. 17-19. Such a device uses the insulating properties of silicon dioxide as the dielectric of the capacitor. One plate of the capacitor is the aluminum metallization, as shown. The other plate is formed from the heavily doped layer just beneath the oxide, labeled N^+. Values of capacitance up to 500 pF can be formed, with maximum voltages of about 50.

An alternate method uses tantalum oxide as the dielectric, producing values up to 5000 pF with working voltages of about 20. Several other processes are used to limited degrees, each having certain advantages and disadvantages.

A resistor is formed by simply defining the dimensions of a certain volume of silicon, properly doped, and isolating it from the other components. The dimensions of a typical resistor are shown in Fig. 17-20a, and the actual construction in Fig. 17-20b. Directly beneath the silicon dioxide is the P-type layer that is the resistor proper. Each end connects to the aluminum metallization pattern. Such a

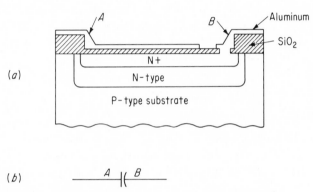

FIG. 17-19.   (a) Silicon dioxide capacitor used in monolithic ICs; (b) schematic representation.

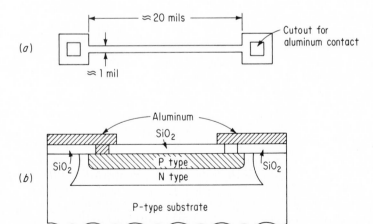

FIG. 17-20.   (a) Approximate dimensions of a diffused resistor; (b) internal resistor structure.

unit, with these dimensions and with typical diffusion densities, will have a resistance of approximately 4000 Ω. In considering the entire integrated chip, the resistors are usually formed at the same time as the transistor bases, and the value of each is determined by the dimensions of the pattern used. Note in Fig. 17-20 that there is a parasitic PNP transistor formed. For this reason, it is necessary to ensure that the proper voltages are provided to keep the junctions reverse-biased at all times. This is accomplished by returning the N-type layer just beneath the P-type resistor to the most positive voltage available. Thus, the PNP transistor can never become turned on.

The foregoing description has been greatly simplified and should not be considered to be a complete dissertation on the subject. Various manufacturers have differing processes, not all of which have been taken into account. Nevertheless, this discussion is adequate for allowing one to most efficaciously apply ICs.

17-3 THE OPERATIONAL AMPLIFIER

The operational amplifier is a prime example of a linear integrated circuit. Recall that in Chap. 14 this circuit configuration was touched upon, with the implication that the circuit was implemented by the use of discrete techniques. Here, we shall use as an example the

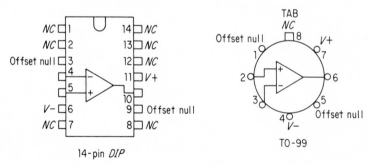

FIG. 17-21.   μA741C operational-amplifier pin connections.

μA741C Operational Amplifier. Pin connections for two different cases are illustrated in Fig. 17-21.

The characteristics of this circuit are such that the open-loop voltage gain is typically 100,000. This is certainly high enough for use in an operational-amplifier configuration. Other characteristics of importance to us are:

Maximum supply voltage	± 18 V
Total power dissipation	500 mW
Maximum input difference voltage	± 30 V
Operating temperature range	0 to 70°C

This particular chip is a high performance monolithic (single-stone) operational amplifier constructed on a single silicon chip. It

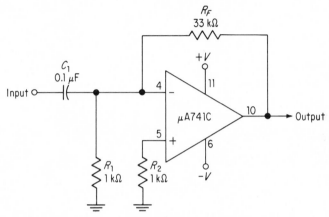

FIG. 17-22.   Operational-amplifier circuit.

contains 21 transistors and 1 diode, along with 11 resistors and 1 capacitor. Some of its intended applications include use as an integrator, summing amplifier, or any of the several feedback configurations. Its internal 6-dB rolloff ensures stability in closed-loop applications.

A practical circuit application for this IC is shown in Fig. 17-22. This operational amplifier has a voltage gain of approximately 30, depending to some degree upon the source impedance. As shown, without considering the source impedance, the voltage gain is determined by $R_F/R_1 = 33,000/1000 = 33$. However, the source impedance must be considered for a true value. As mentioned in Chap. 14, the voltage gain, as well as other circuit parameters, will remain constant over the life of the equipment, which is one of the major advantages of the operational amplifier. While it is of little advantage to know what the IC consists of internally, Fig. 17-23 illustrates the schematic diagram for this unit. Note the complexity compared to a more ordinary discrete-transistor variety. A number of integrated-circuit techniques are illustrated in this schematic diagram. Note the capacitor C_1: this is evidence that it is possible to form capacitors using

IC processes, as described earlier. Also, note Q_{11}. This is a transistor that, by connecting the collector to the base, has been altered to perform as a diode. Again, this is common practice, where the base-emitter junction is used by itself as a diode. It might be mentioned in passing that such an arrangement can also be used as a zener diode, using the reverse breakdown voltage of the emitter-base junction to provide the zener voltage. Of course, the circuit must provide ade-

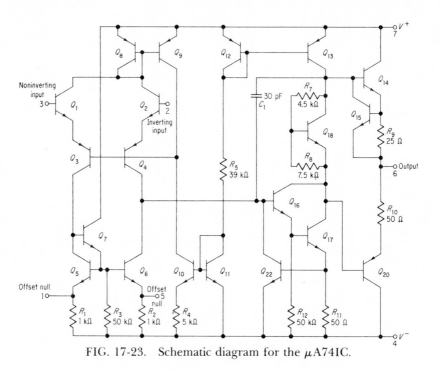

FIG. 17-23.   Schematic diagram for the μA74IC.

quate current limiting to prevent overloading the junction. Section 17-4 describes such an instance.

17-4 NONSATURATING LIMITING AMPLIFIER

The circuit illustrated in Fig. 17-24 has been chosen to illustrate several applications of integrated circuits. Shown are a Darlington-connected pair, a constant-current source, and a transistor used as a zener diode. The circuit configuration shown in Fig. 17-24 is difficult to read, since the individual components are not drawn according to their circuit function. The circuit is redrawn in Fig. 17-25 to more clearly indicate the purpose of each part. Q_1 and Q_2 are operating in a conventional single-ended differential amplifier circuit. Q_4 is acting as a constant-current source for the differential amplifier, to ensure a very high impedance in the emitter circuit and to provide a current source that cannot be exceeded. This, of course, provides current limiting. Q_3 is operating as a zener diode, with its emitter-base junction operating in the avalanche region, but with current limited by the

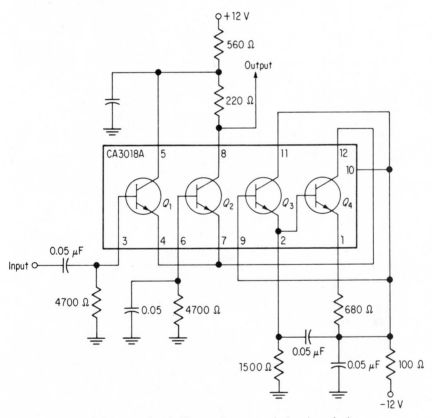

FIG. 17-24. I_C circuit illustrating several circuit techniques.

1500-Ω resistor in the emitter lead, and by the 100-Ω resistor in the collector path.

This particular circuit is used in an application that requires the output amplitude to be absolutely constant. The signal is a sine wave at 3.58 MHz. The voltage gain is essentially unity, and the differential amplifier with the constant-current source in the emitter circuit provides current limitation in one direction. In practice, the output of this circuit is fed to a second circuit identical to this one for limitation in the other direction.

Describing the circuit action in further detail, note that both transistors (Q_1 and Q_2) have the same value of resistance in each base lead. This simply ensures that the bases will be at the same quiescent voltage. The input signal is applied to the base of Q_1, which acts as an

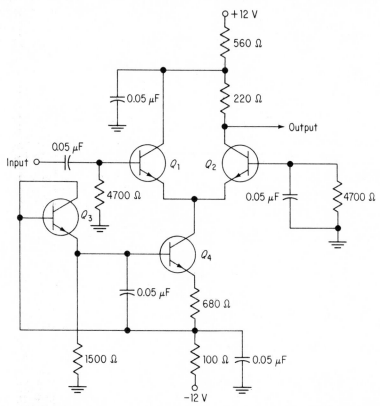

FIG. 17-25.   Circuit of Fig. 17-24 redrawn to emphasize transistor function.

emitter-follower, driving the emitter of Q_2. Q_2 is essentially a common-base circuit, with the output taken from the collector. The 220-Ω resistor is the load resistor, and its value is low because the voltage gain for this particular application is unity. Q_1 and Q_2, then, are seen to be in a conventional differential-amplifier configuration with single-ended input and output.

As mentioned, Q_4 and associated circuitry form a constant-current source. The base voltage is held very constant by means of a zener diode (Q_3), and this causes the collector current to be equally constant. Using a transistor as a zener diode is fairly commonplace. If the base-emitter breakdown voltage is suitable for the application, one needs merely to ensure that the total current is limited to a value less than that which will cause overheating. In this instance, current through the transistor is limited to approximately 10 mA. Figure 17-26 is offered to help in visualizing the fact that Q_3 is operating as a zener diode.

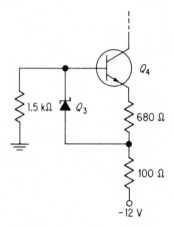

FIG. 17-26. Transistor used as a zener diode.

QUESTIONS AND PROBLEMS

17-1 Briefly define *monolithic*.

17-2 Briefly define *epitaxial layer*.

17-3 Briefly define *microelectronics*.

17-4 Briefly define *hybrid circuit*.

17-5 Briefly define *compatible circuit*.

17-6 Briefly describe a *parasitic transistor*.

17-7 Briefly describe how a parasitic transistor may be made ineffectual.

17-8 What is the primary advantage of an operational amplifier?

17-9 Refer to Fig. 17-22. Which component(s) would be changed to alter the voltage gain?

17-10 Refer to Fig. 17-25. (*a*) Briefly describe what Q_3 is functioning as. (*b*) Explain how this is possible.

GLOSSARY

A_i	Circuit current gain
A_v	Circuit voltage gain
ERCA	Equivalent resistance circuit analysis
h_{FE}	DC current gain (beta)
h_{fe}	AC, or signal, current gain (β_{ac})
I_B	DC base current
I_C	DC collector current
I_{cbo}	Collector-base leakage current
I_{co}	Diode leakage current
I_D	Diode current
I_E	DC emitter current
Q	Quiescent operating point
P_D	Diode power dissipation
RB	Base bias resistor through which base current flows
RB'	Base bias resistor through which base current does not flow
R_{BE}	Base-emitter resistance, viewed from the base $(\beta + 1)R_{EE}$
R_{CB}	Collector-to-base resistance, approximately equal to V_{CE}/I_C
r_{df}	Diode dynamic forward resistance
RE	Resistance in the emitter leg
R_{EE}	Ohmic, or bulk, dc resistance of the emitter material; $100/I_E(\text{ma})\ \Omega$
r_e	Resistance of emitter to signal (ac) current, $26/I_E(\text{ma})\ \Omega$
R_f	Diode forward resistance
R_{ib}	Base to ground resistance $(\beta + 1)(R_{EE} + RE)$
r_i	Total signal input resistance (impedance); also the parallel resistance of RB, RB' and R_{ib}
RL	Resistance in the collector leg; load resistor
r_l	Total signal load
R_R	Diode reverse resistance
R_2	Diode zener resistance, dc
r_z	Diode zener resistance, ac
V_B	Base-to-ground voltage
V_{BB}	Base supply voltage
V_{BE}	Base-to-emitter voltage drop
V_C	Collector-to-ground voltage
V_{CB}	Collector-to-base voltage
V_{CC}	Collector supply voltage
V_{CE}	Collector-to-emitter voltage
V_D	Diode voltage drop
V_E	Emitter-to-ground voltage
V_{EE}	Emitter supply voltage
α	Alpha: emitter-to-collector current gain
β	Beta: base-to-collector current gain, h_{FC} or h_{fe}
γ	Gamma: ratio of V_{CE} to V_{CC}; also, $R_{CB}/(RL + R_{CB} + RE)$

SELECTED READINGS

Boylestad, R., and L. Nashelsky: "Electronic Devices and Circuit Theory," Prentice-Hall, Inc., Englewood Cliffs, N.J., 1972.

Cowles, L. G., "Transistor Circuits and Applications, 2d ed., Prentice-Hall, Inc., Englewood Cliffs, N.J., 1974.

Deboo, G. J., and C. N. Burrous: "Integrated Circuits and Semiconductor Devices, 2d ed.," McGraw-Hill Book Company, New York, 1977.

Kiver, M. S.: "Transistor and Integrated Electronics," 4th ed., McGraw-Hill Book Company, New York, 1972.

Lurch, E. N.: "Fundamentals of Electronics," 2d ed., John Wiley & Sons, Inc., New York, 1971.

Malvino, A. P.: "Electronic Principles," McGraw-Hill Book Company, New York, 1973.

Malvino, A. P.: "Transistor Circuit Approximations," 2d ed., McGraw-Hill Book Company, New York, 1973.

Motorola: "The Semiconductor Data Book," 5th ed., Motorola Semiconductor Products, Inc., Phoenix, Az., 1970.

New York Institute of Technology, "A Programmed Course in Basic Electronics," 2d ed., McGraw-Hill Book Company, New York, 1976.

RCA: "Thyristors, Rectifiers, and Other Diodes," RCA Solid-State Division, Somerville, N.J., 1972.

Richman, P., "MOS Field-Effect Transistors and Integrated Circuits," John Wiley & Sons, Inc., New York, 1974.

Tocci, R. J., "Fundamentals of Electronic Devices," Merrill Publishing Company, Columbus, O., 1970.

ANSWERS
TO SELECTED
PROBLEMS

Chapter 1

1-1	True
1-3	False
1-5	False
1-7	False
1-9	True
1-11	True
1-13	False

Chapter 2

2-1	True
2-3	True
2-5	False
2-7	True
2-9	False
2-11	False
2-13	$I_D = 0.012$ A
2-15	(a) $I_D = 0.0117$ A
	(b) $R_F = 25.6$ Ω
2-17	$R_R = 10^7$ Ω
2-19	$r_{df} = 8.67$ Ω
2-21	$R_Z = 667$ Ω
2-23	$r_z = 10$ Ω

Chapter 3

3-1	(a)
3-3	(c)
3-5	(d)
3-7	(c)
3-9	$\alpha = 0.95$
	$\beta = 19$
3-11	$I_C = 0.000323$ A $= 0.323$ mA
3-13	β is highest when emitter current is 2.02 mA.

Chapter 4

4-1 (c)
4-3 (d)
4-5 False
4-7 (c)
4-9 7.0 mA
4-11 (c)
4-13 (a)
4-15 (a)
4-17 (a)

Chapter 5

5-1 ≈ 0 V
5-3 11.2 kΩ
5-5 I_{RB} Magnitude is determined by RB and RB', usually about 10 times I_E.
  I_B Magnitude is determined by RB and RE; $I_E/(\beta+1)$.
  I_C Magnitude is determined by I_B; $I_B \times \beta$.
  I_E Magnitude is determined by I_B; $I_B \times (\beta+1)$.
  I_{cbo} Magnitude is determined by the temperature.
5-7 A voltage at the base that increases in a direction to increase forward bias causes the base voltage to move away from ground, and so more collector current flows and the collector voltage falls toward ground. In either the NPN or PNP case, negative input yields a positive output, while positive input yields a negative output.
5-9 Alternative paths exist for I_s, which must flow in RB and RB'.
5-11 $V_B \cong -1.7$ V
  $V_E \cong -1.7$ V
  $E_{RL} \cong 8.5$ V
  $I_E \cong 1.7$ mA
5-13 $\cong 58.8$ Ω
5-15 $V_B \cong -2.9$ V
  $V_E \cong -2.9$ V
  $E_{RL} \cong 14.5$ V
  $I_E \cong 2.9$ mA
5-17 $\Delta I_C = 1.1$ mA; $\Delta V_{CE} = 1.4$ V
5-19 2.2 V
5-21 364 kΩ if V_{BE} is ignored; 395 kΩ if V_{BE} is considered.

5-23 (a)

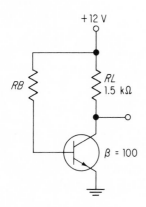

(b) 281 kΩ

(c) 4 mA

(d) 40 μA

5-25   131 kΩ

5-27   1275 Ω

Chapter 6

6-1 $\gamma = 0.25$

6-3 $\gamma = 0.58$; $RB = 75$ kΩ; $RB' = 15$ kΩ; $RL = 7.5$ kΩ; $RE = 5$ kΩ; $R_{CB} = 17.5$ kΩ

6-5   To provide a path for signal current around RE. Makes a virtual dc source at signal frequencies.

6-7 $R_{ib} = 205$ kΩ

6-9 $A_v \cong 5$

6-11 $V_B = 1.67$ V

$V_C = 6.7$ V

$V_{CE} = 5$ V

$\gamma = 0.5$

6-13 $-V_{CC}$ is smaller, and maximum output voltage is determined by V_{CC}.

6-15 $A_v = 96$

6-17 $V_E = 0.39$ V

6-19 $\gamma = 0.82$ V

Chapter 7

7-1   False

7-3   True

7-5   (a) 3
          (b) 4
          (c) 1
7-7 $r_{i(1)} = 470$ kΩ
          $r_{i(2)} = 235$ kΩ
          The input resistance is directly a function of the beta of the transistor; as beta decreases, R_{ib} decreases; as beta increases, R_{ib} increases.
7-9 $r_i = 2.15$ kΩ
7-11 (d)
7-13 (c)
7-15 In the common-collector circuit the collector is at signal ground, which is common to both the input and output.
7-17 $\gamma = 0.75$
7-19 $I_E = 8$ mA
7-21 $A_P = 101$
7-23 $R_{ib} = 101$ kΩ
7-25 $RE = 667$ Ω
7-27 $RB' = 15$ kΩ

Chapter 8

8-1   (c)
8-3   True
8-5 $V_C = -4.57$ V
8-7   True
8-9 $V_C \cong 0$ V
          $V_B \cong 0$ V
          $V_E \cong 0$ V
          $I_C \cong 2.4$ mA
          (Transistor is saturated.)
8-11 $V_B = 2.3$ V
          $V_E = 2.1$ V
          $V_C = 10.79 \cong 10.8$ V
8-13 $A_v = 12.5$
8-15 2500 Ω
8-17 1.06 mV
8-19 32 μF

Chapter 9

9-1 $\gamma = 0.5$
          Q is center of dc load line.

9-3   (c)

9-5 $R_{CB} = 8 \ \text{k}\Omega$

9-7 $RL \cong 20.8 \ \text{k}\Omega$

9-9 $R_{CB} = 10 \ \text{k}\Omega$
          $V_C = -3.5 \ \text{V}$
          $I_C = 325 \ \mu\text{A}$
          $I_B = 3.25 \ \mu\text{A}$

9-11   800 Ω
          -4 V
          0.4
          5 mA
          50 μA

9-13   3300 Ω
          1.71 mA
          22 μA
          5.73 V
          0.57

9-15   5850 Ω
          1.68 mA
          13 μA
          9.87 V
          0.49

Chapter 10

10-1 $V_{CE(\text{max})} = 6.6 \ \text{V}$
          $I_{C(\text{max})} = 9.9 \ \text{mA}$

10-3   Changing RB' to a lower value will result in a greater $V_{CE(\text{max})}$ for signal conditions.

10-5 $I_{RL(\text{min})}$ results from the charge on the coupling capacitor. It flows, even though Q_1 is cut off momentarily, causing a drop across RL.

10-7   Before a signal is applied, the transistor *is* operating at the Q point. Hence, any deviation from this caused by the application of a symmetrical input signal must start from the operating point.

10-9   18 V
          4.5 mA
          10.8 V
          7.3 mA

10-11   5.8 mA

10-13 C_{F1} reduces or eliminates negative feedback, which in turn will reduce the gain and decrease the input impedance.

10-15 $\cong 0$ V
            5.82 mA
            < 0.1 (saturated)

Note: Make certain to allow some leeway in the answers, since some are to be read from the graph, with attendant reading errors.

Chapter 11

11-1   One purpose of the biasing network is to provide the proper operating conditions for the transistor (V_{CE}, I_C, I_B). The other purpose is to provide temperature stability.

11-3   (e)
11-5   (b)
11-7   (d)
11-9   (b)
11-11   (d)
11-13   Total leakage $= (\beta + 1)\, I_{cbo}$

$$I_{cbo} = \frac{\text{total leakage}}{\beta + 1} = \frac{500\ \mu\text{A}}{124 + 1} = 4\ \mu\text{A}$$

11-15 $16\ \mu$A
11-17 $48\ \mu$A
11-19 $256\ \mu$A
11-21 $205\ \mu$A
11-23   20 mW

Chapter 12

12-1 $A_v = 385$
            $r_i = 2.15$ kΩ
            $\gamma = 0.5$
12-3   (a) $V_{C(\text{max})} = -80$ V
                  $V_{C(\text{min})} = 0$ V
            (b) $e_{\text{in}} = 1.47$ V, peak to peak
12-5 $r_i = 3.5$ kΩ
12-7 $A_v = 12$
12-9 $A_v = 27.6$
12-11   17.5 kΩ
12-13   17.5 V
12-15   0.875
12-17   96

12-19	$2000 \ \Omega$
12-21	$15 \ V$
12-23	0.5
12-25	4.0
12-27	$2000 \ \Omega$
12-29	$15 \ V$
12-31	0.5
12-33	$9000 \ \Omega$

Chapter 13

13-1	(c)
13-3	$692 \ kHz$
13-5	$4023 \ Hz$
13-7	Off
13-9	$0.74 \ \mu H$

Chapter 14

14-1	(c)
14-3	(a)
14-5	(d)
14-7	The bias is derived partly from the AVC line. This increases the bias for strong signals and reduces it for weak signals.
14-9	L_2 is a transformer the secondary of which is tuned and forms the resonant circuit for the local oscillator. Energy is fed from the collector of Q_1 to its emitter to produce sustained oscillations.

Chapter 15

15-1	$(a) -$	$(b) +$	$(c) -$
15-3	(a)		
15-5	(d)		
15-7	(a)		
15-9	$(a) \ 4$	$(b) \ 1$	$(c) \ 4$
15-11	$(a) \ 1$	$(b) \ 3$	$(c) \ 5$

Chapter 16

16-1	(a)
16-3	False

16-5   A depletion region always forms at an NP junction, because of recombination forces.

16-7   False

Chapter 17

17-1   Single-stone

17-3   Methods of compressing more components into smaller packages.

17-5   An IC using monolithic techniques for active devices, but passive components are deposited on top of the insulating layer by thin-film techniques.

17-7   By connecting the epitaxial layer to a potential so as to not allow the parasitic unit to become turned on.

17-9 R_F or R_1. Also, the source impedance influences the gain by shunting R_1.

INDEX